P9-BYI-176

Statistics
DeMYSTiFieD®

DeMYSTiFieD® Series

Accounting Demystified
Advanced Calculus Demystified
Advanced Physics Demystified
Advanced Statistics Demystified
Algebra Demystified
Alternative Energy Demystified
Anatomy Demystified
asp.net 2.0 Demystified
Astronomy Demystified
Audio Demystified
Biology Demystified
Biotechnology Demystified
Business Calculus Demystified
Business Math Demystified
Business Statistics Demystified
C++ Demystified
Calculus Demystified
Chemistry Demystified
Circuit Analysis Demystified
College Algebra Demystified
Corporate Finance Demystified
Databases Demystified
Data Structures Demystified
Differential Equations Demystified
Digital Electronics Demystified
Earth Science Demystified
Electricity Demystified
Electronics Demystified
Engineering Statistics Demystified
Environmental Science Demystified
Everyday Math Demystified
Fertility Demystified
Financial Planning Demystified
Forensics Demystified
French Demystified
Genetics Demystified
Geometry Demystified
German Demystified
Home Networking Demystified
Investing Demystified
Italian Demystified
Java Demystified
JavaScript Demystified
Lean Six Sigma Demystified
Linear Algebra Demystified

Logic Demystified
Macroeconomics Demystified
Management Accounting Demystified
Math Proofs Demystified
Math Word Problems Demystified
MATLAB ® Demystified
Medical Billing and Coding Demystified
Medical Terminology Demystified
Meteorology Demystified
Microbiology Demystified
Microeconomics Demystified
Nanotechnology Demystified
Nurse Management Demystified
OOP Demystified
Options Demystified
Organic Chemistry Demystified
Personal Computing Demystified
Pharmacology Demystified
Physics Demystified
Physiology Demystified
Pre-Algebra Demystified
Precalculus Demystified
Probability Demystified
Project Management Demystified
Psychology Demystified
Quality Management Demystified
Quantum Mechanics Demystified
Real Estate Math Demystified
Relativity Demystified
Robotics Demystified
Sales Management Demystified
Signals and Systems Demystified
Six Sigma Demystified
Spanish Demystified
sql Demystified
Statics and Dynamics Demystified
Statistics Demystified
Technical Analysis Demystified
Technical Math Demystified
Trigonometry Demystified
uml Demystified
Visual Basic 2005 Demystified
Visual C# 2005 Demystified
xml Demystified

Statistics

DeMYSTiFieD®

Stan Gibilisco

Second Edition

BROWNSBURG PUBLIC LIBRARY
450 South Jefferson Street
Brownsburg, IN 46112

New York   Chicago   San Francisco   Lisbon   London   Madrid   Mexico City
Milan   New Delhi   San Juan   Seoul   Singapore   Sydney   Toronto

The McGraw-Hill Companies

Cataloging-in-Publication Data is on file with the Library of Congress

McGraw-Hill books are available at special quantity discounts to use as premiums and sales promotions, or for use in corporate training programs. To contact a representative please e-mail us at bulksales@mcgraw-hill.com.

Statistics DeMYSTiFieD®, Second Edition

Copyright © 2011 by The McGraw-Hill Companies, Inc. All rights reserved. Printed in the United States of America. Except as permitted under the United States Copyright Act of 1976, no part of this publication may be reproduced or distributed in any form or by any means, or stored in a data base or retrieval system, without the prior written permission of the publisher.

1 2 3 4 5 6 7 8 9 0   DOC/DOC   1 9 8 7 6 5 4 3 2 1

ISBN 978-0-07-175133-9
MHID 0-07-175133-5

Sponsoring Editor	**Proofreader**
Judy Bass	Surendra Nath Shivam,
Acquisitions Coordinator	Glyph International
Michael Mulcahy	**Production Supervisor**
Editing Supervisor	Pamela A. Pelton
David E. Fogarty	**Composition**
Project Managers	Glyph International
Sapna Rastogi and Tania Andrabi,	**Art Director, Cover**
Glyph International	Jeff Weeks
Copy Editor	**Cover Illustration**
Manish Tiwari,	Lance Lekander
Glyph International	

Trademarks: McGraw-Hill, the McGraw-Hill Publishing logo, Demystified, and related trade dress are trademarks or registered trademarks of The McGraw-Hill Companies and/or its affiliates in the United States and other countries and may not be used without written permission. All other trademarks are the property of their respective owners. The McGraw-Hill Companies is not associated with any product or vendor mentioned in this book.

Information contained in this work has been obtained by The McGraw-Hill Companies, Inc. ("McGraw-Hill") from sources believed to be reliable. However, neither McGraw-Hill nor its authors guarantee the accuracy or completeness of any information published herein, and neither McGraw-Hill nor its authors shall be responsible for any errors, omissions, or damages arising out of use of this information. This work is published with the understanding that McGraw-Hill and its authors are supplying information but are not attempting to render engineering or other professional services. If such services are required, the assistance of an appropriate professional should be sought.

To Tony, Tim, and Samuel
from Uncle Stan

About the Author

Stan Gibilisco, an electronics engineer, researcher, and mathematician, has authored multiple titles for the McGraw-Hill *Demystified* series, along with numerous other technical books and dozens of magazine articles. His work has been published in several languages.

Contents

Acknowledgments

I extend thanks to my nephew Tony Boutelle, a technical writer who lives in Minneapolis. He spent many hours helping me proofread the manuscript, and he offered insights and suggestions from the viewpoint of the intended audience.

How to Use This Book

This book can help you learn basic statistics without taking a formal course. It can also serve as a supplemental text in a classroom, tutored, or home-schooling environment. I recommend that you start at the beginning of this book and go straight through.

In order to learn statistics, you *must* have some mathematical skill. If I told you otherwise, I'd be cheating you. None of the mathematics in this book goes beyond the high-school level. If you need a refresher, you can select from several *Demystified* books dedicated to mathematics topics. If you want to build yourself a "rock-solid" mathematics foundation before you start this course, I recommend that you go through *Algebra Know-It-All* and *Pre-Calculus Know-It-All*.

This book contains abundant multiple-choice questions written in standardized test format. You'll find an "open-book" quiz at the end of every chapter. You may (and should!) refer to the chapter texts when taking these quizzes. Write down your answers, and then give your list of answers to a friend. Have your friend tell you your score, but not which questions you missed. The correct answers appear in the back of the book. Stick with a chapter until you get most of the quiz answers correct.

Two major sections constitute this course. Each section ends with a multiple-choice test. Take these tests when you're done with the respective sections and have taken all the chapter quizzes. Don't look back at the text when taking the section tests. They're easier than the chapter-ending quizzes, and they don't require you to memorize trivial things. A satisfactory score is three-quarters correct. Answers appear in the back of the book.

The course concludes with a 100-question final exam. Take it when you've finished all the sections, all the section tests, and all of the chapter quizzes. A satisfactory score is at least 75 percent correct answers.

With the section tests and the final exam, as with the quizzes, have a friend divulge your score without letting you know which questions you missed. That way, you won't subconsciously memorize the answers. You might want to take each test, and the final exam, two or three times. When you get a score that makes you happy, you can (and should!) check to see where your strengths and weaknesses lie.

You should strive to complete one chapter of this book every 10 days or two weeks. A couple of hours daily ought to prove sufficient for this task. Don't rush yourself. Give your mind time to absorb the material. But don't go too slowly either. Proceed at a steady pace and keep it up. That way, you'll complete the course in a few months. (As much as we all wish otherwise, nothing can substitute for "good study habits.") When you're done with the course, you can use this book as a permanent reference.

I welcome your ideas and suggestions for future editions.

Stan Gibilisco

Statistics
DeMYSTiFieD®

Part I

Statistics Concepts

chapter 1

Background Mathematics

This chapter offers a review of some basic mathematics principles. When you want to know what you're reading, writing, or talking about in statistics, you must have some familiarity with basic mathematics, including set theory, number theory, relations, functions, equations, and graphs. Table 1-1 lists common symbols used in mathematics. You'll encounter many of these symbols in statistics as well.

CHAPTER OBJECTIVES

In this chapter, you will

- Define, combine, and compare sets
- Learn how variables affect each other
- Distinguish between mathematical relations and functions
- Discover how relations and functions operate
- Construct and define sets of numbers
- Solve and graph simple algebraic equations

TABLE 1-1   Symbols used in mathematics.

Symbol	Description
{ }	Braces; objects between them are elements of a set
⇒	Logical implication; read "implies"
⇔	Logical equivalence; read "if and only if"
∀	Universal quantifier; read "for all" or "for every"
∃	Existential quantifier; read "for some"
| :	Logical expression; read "such that"
&	Logical conjunction; read "and"
N	The set of natural numbers
Z	The set of integers
Q	The set of rational numbers
R	The set of real numbers
ℵ	Transfinite (infinite) cardinal number
∅	The set with no elements; read "the empty set" or "the null set"
∩	Set intersection; read "intersect"
∪	Set union; read "union"
⊂	Proper subset; read "is a proper subset of"
⊆	Subset; read "is a subset of"
∈	Element; read "is an element of" or "is a member of"
∉	Nonelement; read "is not an element of" or "is not a member of"
=	Equality; read "equals" or "is equal to"
≠	Inequality; read "does not equal" or "is not equal to"
≈	Approximate equality; read "is approximately equal to"
<	Inequality; read "is less than"
≤	Equality or inequality; read "is less than or equal to"
>	Inequality; read "is greater than"
≥	Equality or inequality; read "is greater than or equal to"
+	Addition; read "plus"
−	Subtraction, read "minus"
× · *	Multiplication; read "times" or "multiplied by"
÷ /	Quotient; read "over" or "divided by"

(Continued)

TABLE 1-1 Symbols used in mathematics. (*Continued*)	
Symbol	**Description**
:	Ratio or proportion; read "is to"
	Logical expression; read "it is true that"
!	Product of all natural numbers from 1 up to a certain value; read "factorial"
( )	Quantification; read "the quantity"
[ ]	Quantification; used outside ( )
{ }	Quantification; used outside [ ]

Sets

A *set* constitutes a collection or group of definable things called *elements* or *members*. Examples of sets include the following:

- Points on a line
- Points in a plane
- Points in space
- Points (or moments) in time
- Apples in a basket
- Coordinates on a display
- Curves and lines on a graph
- Chemical elements
- People in a city
- Locations in computer memory
- Data bits on a computer disk
- Subscribers to an Internet blog

If an object or number (let's call it a) constitutes an element of set A, we can denote the fact by writing

$$a \in A$$

The $\in$ symbol means "is an element of." We can also say that "the element a belongs to the set A" or that "the set A contains the element a." If an object b does not constitute an element of set A, we can denote that fact by writing

$$b \notin A$$

Set Intersection

We define the *intersection* of two sets A and B, written $A \cap B$, as a third set C such that the following statement holds true for every possible element x:

$$x \in C \text{ if and only if } x \in A \text{ and } x \in B$$

We can read the $\cap$ symbol as "intersect."

Set Union

We define the *union* of two sets A and B, written $A \cup B$, as a third set C such that the following statement holds true for every possible element x:

$$x \in C \text{ if and only if } x \in A \text{ or } x \in B$$

We can read the $\cup$ symbol as "union."

Subsets

A set A constitutes a *subset* of a set B, written $A \subseteq B$, if and only if the following statement holds true for every possible element x:

$$x \in A \text{ implies that } x \in B$$

We read the symbol $\subseteq$ out loud as "is a subset of." In this context, the phrase "implies that" operates in the strongest possible sense. Logically, "implies" doesn't mean "makes likely" or "strongly suggests." It's a mandate! Therefore, we can also write the above statement as

$$\text{If } x \in A, \text{ then } x \in B$$

Proper Subsets

A set A constitutes a *proper subset* of a set B, written $A \subset B$, if and only if both of the following statements hold true for every possible element x:

$$x \in A \text{ implies that } x \in B$$

and

$$A \neq B$$

We read the symbol $\subset$ as "is a proper subset of."

Disjoint Sets

We define two sets A and B as *disjoint sets* if and only if their intersection contains no elements; that is, the following condition holds true:

$$A \cap B = \varnothing$$

where $\varnothing$ denotes the *empty set*, also called the *null set*. The empty set doesn't contain any elements, but it still exists, like a basket of apples without the apples.

Coincident Sets

We define two nonempty sets A and B as *coincident sets* if and only if, for all possible elements x, both of the following statements hold true:

$$x \in A \text{ implies that } x \in B$$

and

$$x \in B \text{ implies that } x \in A$$

When two sets A and B coincide, we write

$$A = B$$

？ Still Struggling

Do you wonder what makes a collection of things into a set, and not just "the things themselves"? If you have a basket full of apples and you call it a set, does it still form a set when you dump the apples onto the ground? Did those same apples form the elements of a set before someone picked them off the trees? The answers to these questions are up to you. A collection of objects forms a set *if and only if* you decide to call it a set.

Here's a Riddle!

Can any set ever constitute an element of itself? At first, you might say "No, that's impossible. That would be like saying that the Pingoville Ping-Pong Club is one of its own members. The elements are the Ping-Pong players themselves, not the club as a whole." But wait! What about the set of all abstract ideas? That's an abstract idea, so in some situations a set *can* behave as a member of itself.

Relations and Functions

Consider the following three statements, each of which represents a situation that can occur in everyday life.

- The outdoor air temperature varies with the time of day.
- The length of time that the sun remains above the horizon on June 21 varies with the latitude of the observer.
- The time required for a wet rag to dry depends on the air temperature.

All of these expressions involve some phenomenon that *depends* on something else. In the first case, we make a claim concerning temperature versus time; in the second case, we make a claim concerning sunup time versus latitude; in the third case, we make a claim concerning time versus temperature. In these situations, the term *versus* means "compared with" or "with respect to."

Independent Variables

An *independent variable* changes, but its value does not depend on anything else in a given scenario. Scientists, statisticians, mathematicians, and other technical people commonly treat time as an independent variable, and for good reason: Lots of things depend on time!

When two or more variables interact, at least one of the variables is independent, but they're not all independent. In the three situations described above, the independent variables are time, latitude, and air temperature, respectively.

Dependent Variables

A *dependent variable* changes, but its value depends on at least one other factor in a situation. In the scenarios described above, the air temperature, the sunup time, and rag-drying time constitute the dependent variables.

When two or more variables interact, at least one of them is dependent, but they can't all be dependent. Some phenomenon that acts as an independent variable in one instance might act as a dependent variable in another case. For example, the air temperature constitutes the dependent variable in the first situation described above, but it constitutes the independent variable in the third situation.

Scenarios Illustrated

The three scenarios described above lend themselves to illustration. Figure 1-1 illustrates hypothetical situations of these sorts.

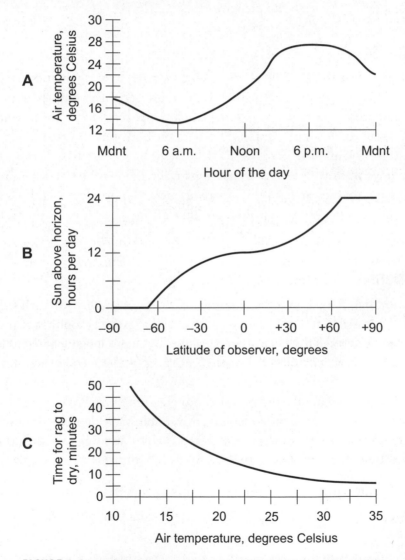

FIGURE 1-1 · Three "this-versus-that" scenarios. At A, air temperature versus time of day; at B, sunup time versus latitude; at C, time for rag to dry versus air temperature.

At Fig. 1-1A, we see an example of outdoor air temperature versus time of day. Figure 1-1B shows the sunup time (the number of hours per day in which the sun is above the horizon) versus latitude on June 21, where points south of the equator have negative latitude and points north of the equator have positive latitude. Figure 1-1C shows the time it takes for a rag to dry, plotted against the air temperature.

The situations represented by Fig. 1-1A and C are pure fiction, having been contrived for this discussion. Figure 1-1B represents true astronomical data for June 21 of every year on the planet earth.

Relations

All three of the graphs in Fig. 1-1 represent *relations*. In mathematics, we define a relation as an expression of the way two or more variables compare or interact. (We could just as well use the terms *relationship*, *comparison*, or *interaction*.) Figure 1-1B, for example, graphs the relation between the latitude and the sunup time on June 21.

When we work with logical or mathematical relations, the statements remain valid if we state the variables the other way around. Therefore, Fig. 1-1B shows a relation between the sunup time on June 21 and the latitude. In a relation, "this versus that" means the same thing as "that versus this."

Functions

A relation describes how the values of variables compare with each other; it works in a "passive" sense. A *function* is a special type of relation that transforms, processes, or morphs the quantity represented by the independent variable into the quantity represented by the dependent variable. A function operates in an "active" sense.

All three of the graphs in Fig. 1-1 represent functions. We can think of the changes in the values of the independent variables as causative factors in the variations of the values of the dependent variables. We might restate the scenarios as follows to emphasize their status as functions:

- The outdoor air temperature constitutes a function of the time of day.
- The sunup time on June 21 constitutes a function of the latitude of the observer.
- The time required for a wet rag to dry constitutes a function of the air temperature.

A relation is a function *if and only if* every element in the set of its independent-variable values has *at most* one correspondent in the set of dependent-variable values. If a given value of the dependent variable in a relation has more than one independent-variable value corresponding to it, then that relation might nevertheless constitute a function. However, if any given value of the independent variable corresponds to more than one dependent-variable value, then that relation is not a function.

Reversing the Variables

In a graph, we'll usually plot the value of the independent variable along (or parallel to) the horizontal axis, and the value of the dependent variable along (or parallel to) the vertical axis. Imagine a movable, vertical line in such a graph. Suppose that we can move that vertical line to the left and right at will. A curve or plot represents a function *if and only if* it never intersects the movable vertical line at more than one point. This so-called *vertical-line test* gives us a quick and easy way to see whether or not a particular graph represents a function of the independent variable.

Suppose that we reverse the independent and dependent variables of the functions shown in Fig. 1-1. This action produces the following strange assertions:

- The time of day constitutes a function of the outdoor air temperature.
- The latitude of an observer constitutes a function of the sunup time on June 21.
- The air temperature constitutes a function of the time it takes for a wet rag to dry.

The first of these statements make no sense. We can't make time go backward, or make time pass at a different rate, by cooling things off or heating things up. The second statement can make sense if we restrict the values to which we apply the function. Our latitude *does* depend on how long the sun remains above the horizon on June 21 (as long as we stay out of the arctic zones where the latitudes exceed 66.5 degrees north or south of the equator; in those zones the sun neither rises nor sets on June 21).

If we turn the graphs of Fig. 1-1A and B sideways to reflect the transposition of the variables and then perform the vertical-line test, we can see that the graphs no longer depict functions if we consider them in their entirety. Therefore, the first of the above assertions is not only absurd, but false. The second of the above assertions makes sense and holds true as long as we restrict our attention to the curved (nonlinear) portion of the graph and eliminate the straight-line (linear) portions.

Figure 1-1C portrays a function, at least in theory, when we "stand it on its side." The statement of the "reverse function" sounds bizarre, but it can hold true under certain conditions. The drying time of a standard-sized wet rag made of a standard material (such as cotton cloth) could be used to infer air temperature experimentally, although the humidity and the wind speed would also have an effect.

Domain and Range

Imagine a mathematical function (let's call it f) that "operates" on elements in a certain set A, "morphing" those elements into members of another set B. Let A' represent the set of all elements a in A for which the function f produces a corresponding element b in B. In this situation, we call set A' the *domain* of the function f.

Now let's go the other way. Once again, suppose that f is a function that takes elements from set A and "morphs" (a mathematician would say *maps*) them into the elements of set B. Let B' represent the set of all elements b in B for which there exists a corresponding element a in A. We call B' the *range* of f.

PROBLEM 1-1

Figure 1-2 is an example of a specialized drawing called a *Venn diagram*. The circles portray two sets A and B. The dots portray three points or elements P, Q, and R. What does the hatched region represent? Which of the three points P, Q, or R, if any, represents an element of $A \cap B$, the intersection of sets A and B? Which of the three points P, Q, or R, if any, represents an element of $A \cup B$, the union of sets A and B?

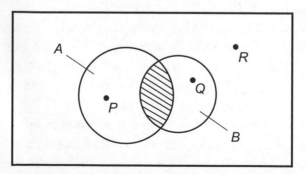

FIGURE 1-2 · Illustration for Problem 1-1.

SOLUTION

The hatched region represents all the elements that belong to both sets *A* and *B*, so it portrays *A* ∩ *B*, the intersection of *A* and *B*. None of the elements (points) shown here lie in the set *A* ∩ *B*. Points *P* and *Q* both lie in *A* ∪ *B*, the union of *A* and *B*. That union set corresponds to the entire region inside one or the other, or both, of the large circles.

PROBLEM 1-2

Figure 1-3 illustrates a relation that maps certain points in a set *C* to certain points in a set *D*. Imagine that only the points shown are involved in this relation. Does the relation constitute a function? If so, how can you tell? If not, why not?

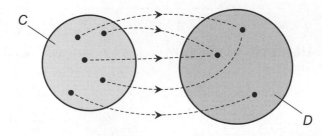

FIGURE 1-3 · Illustration for Problem 1-2.

SOLUTION

The relation is a function, because each value of the independent variable, shown by the points in set *C*, maps into *at most* one value of the dependent variable, represented by the points in set *D*.

Numbers

A *number* is an abstract expression of a quantity. Mathematicians define numbers in terms of a rather esoteric scheme of "sets within sets." Using that methodology, we can build up all of the known numbers from a starting point of zero. We use *numerals* as written symbols to represent numbers.

Natural and Whole Numbers

The *natural numbers*, also called *whole numbers* or *counting numbers*, arise from a starting point of 0 or 1, depending on which text you consult. We can denote the set of natural numbers by writing N. If we include 0, then

$$N = \{0, 1, 2, 3, ..., n, ...\}$$

In some instances, 0 is not included, so we write

$$N = \{1, 2, 3, 4, ..., n, ...\}$$

We can graphically portray natural numbers as individual points that lie along a straight geometric *ray* or *half-line*, where the numerical quantity varies in direct proportion to the displacement (Fig. 1-4). We call such a ray or half-line a *natural-number line*.

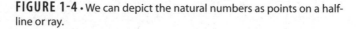

FIGURE 1-4 · We can depict the natural numbers as points on a half-line or ray.

Integers

We can duplicate and invert the set of natural numbers to form the "mirror-image" set of *negatives of the natural numbers*, as follows:

$$-N = \{0, -1, -2, -3, ..., -n, ...\}$$

The union of this set with the set of natural numbers produces the set of *integers*, commonly denoted by Z. Therefore, we can write

$$Z = N \cup -N$$
$$= \{..., -n, ..., -2, -1, 0, 1, 2, ..., n, ...\}$$

We can graphically portray integers as points spaced at equal intervals along a straight line, where the quantity varies in direct proportion to the displacement in either direction (Fig. 1-5). Usually, points for the negative quantities lie to the left of the zero point (or *origin*), and for positive quantities lie to the right of the origin. In this drawing, integers correspond to points where hash marks cross the line.

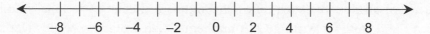

FIGURE 1-5 · We can depict the integers as individual points, spaced at equal intervals on a line.

TIP *The set of natural numbers constitutes a proper subset of the set of integers.
Symbolically, we write this fact as*

$$N \subset Z$$

*For any number a, if a is an element of N, then a is an element of Z. The converse
of this statement, however, does not hold true in general. We can find elements of
Z (namely, the negative integers) that don't belong to N.*

? Still Struggling

The integers can get confusing when you compare values. For example, −5 is
smaller (or less) than −2, and any negative integer is smaller (or less) than any
natural number. Conversely, −2 is larger (or greater) than −5, and any natural
number is larger (or greater) than any negative integer. Does that notion seem
strange? How can −158 be "smaller" than −12? If you find yourself in debt by
$158, don't you have a bigger problem than you'd have if your debt amounted
to only $12? In the literal sense, −158 does indeed represent a smaller numerical
value than −12 does, just as −158 degrees represents a colder temperature than
−12 degrees does. In fact, the integer −158 is less than −12 or −32 or −157. But
−158 is *larger negatively* than −12 or −32 or −157. Let's take a lesson from these
riddles: If we want to avoid confusion when comparing numbers, we had better
choose our words with care!

Rational Numbers

We get a *rational number* (the term derives from the word *ratio*) when we
divide an integer by a positive integer. If *a* represents an integer and *b* represents
a *nonzero natural number* (positive integer), then we can always break down a
rational number *r* into the ratio

$$r = a/b$$

The set of all possible such ratios encompasses the set of rational numbers, which we denote by writing **Q**. A mathematician would write this fact symbolically as

$$\mathbf{Q} = \{x \mid x = a/b\}$$

where $a \in \mathbf{Z}$, $b \in \mathbf{Z}$, and $b > 0$. (We read the x to the left of the vertical line as "All possible elements x." The vertical line means "such that.")

The set of integers is a proper subset of the set of rational numbers. We can now observe that the natural numbers, the integers, and the rational numbers have the following relationship:

$$\mathbf{N} \subset \mathbf{Z} \subset \mathbf{Q}$$

Decimal Expansions

We can denote any rational number in the so-called *decimal form* by writing an integer followed by a period (called a *decimal point* or *radix point*), and then writing a sequence of digits after the period. This digit sequence can take either of two forms:

- A finite string of digits
- An infinite string of digits that repeat with a pattern

Following are three examples of the first type, known as *terminating decimals*:

$$3/4 = 0.750000 \ldots$$

$$-9/8 = -1.1250000 \ldots$$

$$47/1000 = 0.0470000 \ldots$$

Following are three examples of the second type, called *nonterminating, repeating decimals*:

$$1/3 = 0.33333 \ldots$$

$$-123/999 = -0.123123123 \ldots$$

$$1/7 = 0.142857142857142857 \ldots$$

Irrational Numbers

When we can't express a quantity as the ratio of an integer to a positive integer, we call that quantity an *irrational number*. (The term "irrational" in this context means "having no ratio." It doesn't mean "unreasonable" or "insane.") Examples of irrational numbers include:

- The length of the diagonal of a square in a flat plane that measures 1 unit long on each edge; this quantity equals $2^{1/2}$, also known as the square root of 2

- The circumference-to-diameter ratio of a circle as determined in a flat plane, conventionally symbolized as the lowercase Greek letter pi (π)

All irrational numbers share one quirk: We can't precisely express them in decimal form. When we try to do that, we always get a *nonterminating, nonrepeating* decimal. No matter how many digits we specify to the right of the radix point, the expression always remains an approximation of the actual value of the number; we can never "hit" the exact quantity. We can denote decimal approximations by writing down the first few digits to the right of the radix point, followed by an ellipsis (three dots). For example:

$$2^{1/2} = 1.41421356 \ldots$$

$$\pi = 3.14159 \ldots$$

We can use "squiggly equals signs" to indicate that values are approximate, so we could write the above expressions as follows:

$$2^{1/2} \approx 1.41421356$$

$$\pi \approx 3.14159$$

Let's denote the entire set of irrational numbers as S. This set has absolutely no elements in common with the set of rational numbers—even though, in a sense, the two sets "intertwine." Symbolically,

$$Q \cap S = \varnothing$$

Real Numbers

When we take the union of the sets of rational and irrational numbers, we get the set of *real numbers*, denoted by R. Symbolically,

$$R = Q \cup S$$

Mathematicians sometimes depict the set R graphically as a continuous, straight geometric line like the one in Fig. 1-5.

The set of real numbers relates to the sets of rational numbers, integers, and natural numbers as follows:

$$N \subset Z \subset Q \subset R$$

We can define the operations of addition, subtraction, and multiplication over the set of real numbers. If # represents any of these operations and x and y constitute elements of R, then

$$x \# y \in R$$

In the case of division (a quotient or ratio, symbolized as a forward slash), we have

$$x/y \in R$$

if and only if $y \neq 0$.

 PROBLEM 1-3

Given any two different rational numbers, can we always find another rational number between them? That is, if x and y represent two different rational numbers, does some rational number z exist such that $x < z < y$ (x is less than z, and z is less than y)?

SOLUTION

Yes. We can prove this fact. Let's start with the general arithmetic formula for the sum of two fractions:

$$a/b + c/d = (ad + bc)/(bd)$$

where neither b nor d equals 0. Consider two rational numbers x and y comprising ratios of integers $a, b, c,$ and $d,$ such that

$$x = a/b$$

and

$$y = c/d$$

We can find the *arithmetic mean*, also called the *average*, of these two rational numbers; this operation will produce a number z with the characteristics we seek. The arithmetic mean of two numbers equals half the sum of the two numbers. The arithmetic mean of any two rational numbers is always another rational number. We can prove this fact by noting that

$$(x + y)/2 = (a/b + c/d)/2$$

$$= (ad + bc)/(2bd)$$

The product of any two integers equals another integer. Also, the sum of any two integers equals another integer. Therefore, because a, b, c, and d all represent integers, we know that $ad + bc$ is an integer, and also that $2bd$ is an integer. Let's call these derived integers p and q, as follows:

$$p = ad + bc$$

and

$$q = 2bd$$

The arithmetic mean of x and y equals p/q, which constitutes a rational number because it's the ratio of two integers.

Equations with One Variable

When we want to solve a single-variable equation, we must "morph" it into a form where the left-hand side of the equals sign contains *only* the variable whose value we seek (e.g., x), and the right-hand side of the equals sign does *not* contain the variable whose value we seek.

Elementary Rules

We can usually manipulate an equation in one variable using simple algebra to derive a solution—assuming that a solution exists, of course! To accomplish this task, we can take advantage of following rules in any order and as many times as we want.

- Addition of a quantity to each side: Any defined constant, variable, or expression can be added to both sides of an equation, producing a new equation equivalent to the original equation.

- Subtraction of a quantity from each side: Any defined constant, variable, or expression can be subtracted from both sides of an equation, producing a new equation equivalent to the original equation.

- Multiplication of each side by a nonzero quantity: Both sides of an equation can be multiplied by a defined nonzero constant, variable, or expression, producing a new equation equivalent to the original equation.
- Division of each side by a nonzero quantity: Both sides of an equation can be divided by a nonzero constant, by a variable that can never attain a value of zero, or by an expression that can never attain a value of zero, producing a new equation equivalent to the original equation.

Basic Equation in One Variable

Consider an equation of the following form where a, b, c, and d represent real-number constants, x represents a variable, and $a \neq c$:

$$ax + b = cx + d$$

We can solve this equation for x by executing several steps. First, let's subtract b from each side, getting

$$ax = cx + d - b$$

When we subtract cx from each side, we get

$$ax - cx = d - b$$

Separating out the two products on the left-hand side using the distributive principle from basic arithmetic, we obtain

$$(a - c)x = d - b$$

Finally, because $a \neq c$, we can divide the entire equation through by the quantity $(a - c)$ to arrive at

$$x = (d - b)/(a - c)$$

Factored Equations in One Variable

Consider an equation of the following form where a_1, a_2, a_3, ..., and a_n represent real-number constants, and x represents a variable:

$$(x - a_1)(x - a_2)(x - a_3) \ldots (x - a_n) = 0$$

This equation has multiple solutions. We get a solution when any single one of the factors equals zero. That happens when *at least one* of the following conditions is met:

$$x = a_1$$

$$x = a_2$$

$$x = a_3$$

$$\downarrow$$

$$x = a_n$$

If we let A represent the *solution set* of this equation, then

$$A = \{a_1, a_2, a_3, \ldots a_n\}$$

Quadratic Equations

Consider an equation of the following form where a, b, and c represent real-number constants, x represents a variable, and $a \neq 0$:

$$ax^2 + bx + c = 0$$

Mathematicians call this expression the *standard form of a single-variable quadratic equation*. Some such equations have no real-number solutions; others have a single real-number solution; still others possess two real-number solutions. We can find the solution(s) of the above generalized equation (let's call them x_1 and x_2) according to the following formulas:

$$x_1 = [-b + (b^2 - 4ac)^{1/2}]/(2a)$$

and

$$x_2 = [-b - (b^2 - 4ac)^{1/2}]/(2a)$$

Sometimes these formulas appear together as a single formula with a plus-or-minus sign ($\pm$) to indicate that we can apply either addition or subtraction. When we render the solution in that form, we get the *quadratic formula* from elementary algebra:

$$x = [-b \pm (b^2 - 4ac)^{1/2}]/(2a)$$

PROBLEM **1-4**

Find the solution of the following equation:

$$3x - 5 = 2x$$

 SOLUTION

We can put this equation into the form

$$ax + b = cx + d$$

where $a = 3$, $b = -5$, $c = 2$, and $d = 0$. Then, according to the general solution derived above, we obtain

$$
\begin{aligned}
x &= (d - b)/(a - c) \\
&= [0 - (0 - 5)]/(3 - 2) \\
&= 5/1 \\
&= 5
\end{aligned}
$$

PROBLEM **1-5**

Find the real-number solution or solutions of the following equation:

$$x^2 - x = 2$$

SOLUTION

This expression constitutes a single-variable quadratic equation. We can "morph" it into the standard quadratic form by subtracting 2 from each side, obtaining

$$x^2 - x - 2 = 0$$

We can now solve this equation in two different ways. First, let's note that it can be factored to obtain

$$(x + 1)(x - 2) = 0$$

We can see straightaway that two solutions exist: $x_1 = -1$ and $x_2 = 2$. Either of these values of x will satisfy the equation, because they render the first and second terms, respectively, equal to zero.

We can also solve the equation by "brute force" using the quadratic formula, once we have reduced the equation to the standard form. In this case, we have the constants $a = 1$, $b = -1$, and $c = -2$. Restating the quadratic formula, plugging in the values for the constants, and then grinding out the arithmetic, we obtain

$$x = [-b \pm (b^2 - 4ac)^{1/2}]/(2a)$$

$$= \{1 \pm [1^2 - 4 \times 1 \times (-2)]^{1/2}\}/(2 \times 1)$$

$$= \{1 \pm [1 - (-8)]^{1/2}\}/2$$

$$= (1 \pm 9^{1/2})/2$$

$$= (1 \pm 3)/2$$

We've derived two solutions for x, as follows:

$$x_1 = (1 + 3)/2$$

$$= 4/2$$

$$= 2$$

and

$$x_2 = (1 - 3)/2$$

$$= -2/2$$

$$= -1$$

? Still Struggling

These solutions coincide with the ones we got by factoring, as we should expect! It doesn't matter that the results appear in opposite order in the two solution processes.

Simple Graphs

When we can clearly define the variables in a function, or when they can attain only specific values (called *discrete values*), we can render graphs in clear and simple terms.

Smooth Curves

Figure 1-6 shows a graph with two curves, each of which portrays the variations in the price of a hypothetical stock during a business day. Let's call the stocks X and Y. Both curves represent functions of time. You can verify this fact using the vertical-line test. Neither curve intersects a movable, vertical line more than once. Time constitutes the independent variable on this coordinate grid.

Now think of the stock price as the independent variable and time as the dependent variable. To graph the situation in these terms, you can "stand the curves on their sides" as shown in Fig. 1-7. When you use the vertical-line test, you can see that in this interpretation, time behaves like a function of the price of stock X, but time does not act like a function of the price of Stock Y.

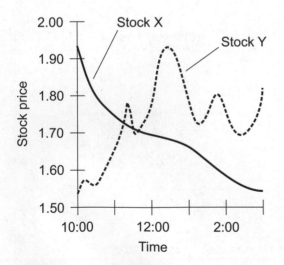

FIGURE 1-6 · The curves show fluctuations in the prices of hypothetical stocks during the course of a business day.

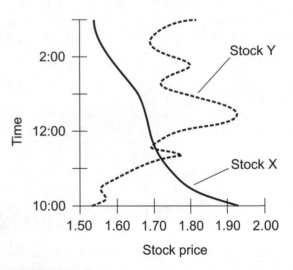

FIGURE 1-7 · A graph in which stock price constitutes the independent variable, and time constitutes the dependent variable.

Vertical Bar Graphs

In a *vertical bar graph*, we portray the independent variable on the horizontal axis and the dependent variable on the vertical axis. We depict function values as the heights of bars having equal widths. Figure 1-8 is a vertical bar graph of the price of the hypothetical stock Y (from the situations of Figs. 1-6 and 1-7) at intervals of 1 hour. This graph reveals less detail than either Fig. 1-6 or 1-7, but some people find this type of graph easier to read.

Horizontal Bar Graphs

In a *horizontal bar graph*, we depict the independent variable on the vertical axis and the dependent variable on the horizontal axis. We plot function values as the widths of bars having equal heights. Figure 1-9 shows a horizontal bar graph of the price of the hypothetical stock Y at intervals of 1 hour. In this example, we go into the future as we move upward along (and parallel to) the vertical axis; we go into the past as we move downward. Some vertical bar graphs reverse the vertical orientation, so that time "flows" from top to bottom.

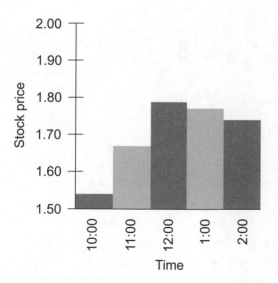

FIGURE 1-8 · Vertical bar graph of hypothetical stock price versus time.

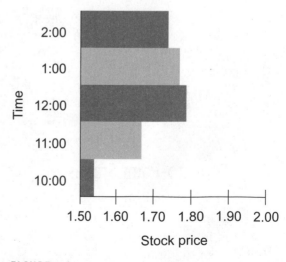

FIGURE 1-9 · Horizontal bar graph of hypothetical stock price versus time.

BROWNSBURG PUBLIC LIBRARY

Histograms

A *histogram* is a bar graph applied to a special situation called a *distribution*. As an example, consider the grades a class receives on a test as shown by Fig. 1-10. Here, each vertical bar represents a letter grade (A, B, C, D, or F). The height of any particular bar represents the percentage of students in the class receiving that grade. This graph shows the values of the dependent variable at the top of each bar. The percentages add up to 100%, based on the assumption that all of the members of the class actually show up, take the test, and turn in their papers.

Some histograms offer greater flexibility than the one in Fig. 1-10, allowing for variable bar widths as well as variable bar heights. We'll see some graphs like this in Chap. 4. Also, in some bar graphs showing percentages, the values don't add up to 100%! We'll see a graph like this later in this chapter.

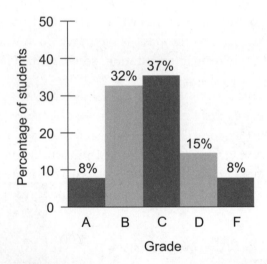

FIGURE 1-10 · A histogram is a specialized form of bar graph.

Point-to-Point Graphs

In a *point-to-point graph*, the scales resemble those used in continuous-curve graphs such as Figs. 1-6 and 1-7, but we show the values only for a few selected points, connecting those points by straight lines rather than a smooth curve.

In the point-to-point graph of Fig. 1-11, we plot the price of stock Y (from the situation we've worked with in this chapter) at every half-hour mark from 10:00 a.m. to 3:00 p.m. The resulting "curve" does not exactly show the stock

BROWNSBURG PUBLIC LIBRARY

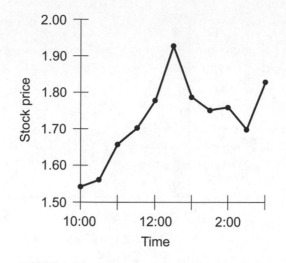

FIGURE 1-11 · A point-to-point graph of hypothetical stock price versus time.

prices at the in-between times. Nevertheless, the graph presents a fairly good representation of the fluctuation of the stock over time.

Whenever we plot a point-to-point graph, we must include a certain minimum number of points and make sure that they exist close enough together. If a point-to-point graph showed the price of stock Y at hourly intervals, it would not come as close as Fig. 1-11 does to representing the actual stock-price function. If a point-to-point graph showed the price at 15-minute intervals, it would come closer than Fig. 1-11 to the moment-to-moment stock-price function.

Don't Confuse the Graph Reader!

Whenever we compose a graph, we should choose sensible scales for the dependent and independent variables. If either scale spans a range of values much greater than necessary, the *resolution* (detail) of the graph will suffer. If either scale does not have a large enough span, then we won't have enough room to show the entire function; some legitimate and important values or points will get "cutoff."

PROBLEM 1-6

Figure 1-12 shows the percentage of the work force in a certain city that "calls in sick" on each day during a particular workweek. What, if anything, is wrong with this graph?

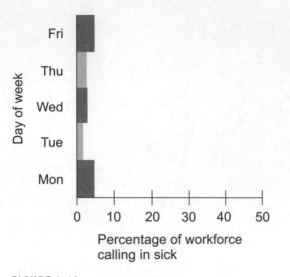

FIGURE 1-12 · Illustration for Problems 1-6 and 1-7.

 SOLUTION

The horizontal scale covers an unnecessarily large range. It makes the values in the graph difficult to ascertain. We would find the graph easier to read, as well as more accurate, if the horizontal scale showed values only in the range of 0 to 10%. We could also improve this graph by writing down the actual percentage numbers at the right-hand side of each bar.

PROBLEM 1-7

What's going on with the percentage values depicted in Fig. 1-12? When we scrutinize this situation, we can see that the values don't add up to 100%. Shouldn't they?

 SOLUTION

Not necessarily. If all the values did add up to 100%, we would have a coincidence—and a bad omen, too, perhaps indicating a disease epidemic! During a bad epidemic, we might see the values add up to more than 100%. If everybody showed up for work every day for the whole week (no one called in sick at all), the sum of the percentages would equal 0%, and the bar graph would appear blank.

Tweaks, Trends, and Correlation

We can approximate or modify graphs by "tweaking" them. We can also use graphs to illustrate certain characteristics of functions, such as trends and correlation.

Linear Interpolation

The term *interpolate* means "to put between." When confronted with an incomplete graph, we can sometimes insert estimated data in the gap(s) to make the graph appear complete. Figure 1-13 shows a graph of the price of our hypothetical stock Y, but a gap occurs during the noon hour. We don't know exactly what happened to the stock price during that hour, but we can fill in the graph using *linear interpolation*. We simply draw a straight line between the end points of the gap.

Linear interpolation almost always produces a somewhat inaccurate result. But sometimes we're better off with an approximation, even a crude one, than we are with no data at all. When you compare Fig. 1-13 with Fig. 1-6, you can see that we have a considerable *linear interpolation error* in this case. Our interpolation completely overlooks the large "hump" from Fig. 1-6, indicating a surge in the stock price of stock Y during the noon hour! Interpolation offers convenience— along with obvious risk.

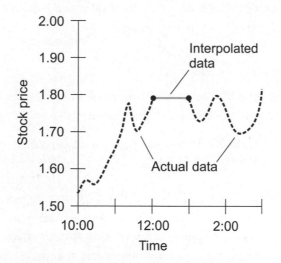

FIGURE 1-13 · An example of linear interpolation. The gray solid line represents the interpolation of the values for the gap in the actual available data (black dashed curve).

? Still Struggling

Can a function exist for which linear interpolation works perfectly, so that the process "fills in the gap" with zero error? The answer is yes. If we know that the graph of a function appears as a straight line, then we can use linear interpolation to "fill in a gap," and the result will have no error. Consider, for example, a car that accelerates at a defined and constant rate. If its speed-versus-time graph appears as a perfectly straight line with a small gap, then we can employ linear interpolation to determine the car's speed at points inside the gap, as shown in Fig. 1-14. In this graph, the black dashed line represents actual measured data, and the gray solid line represents interpolated data.

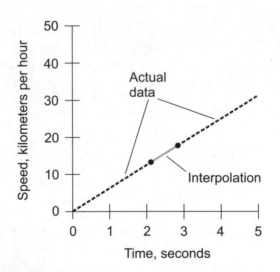

FIGURE 1-14 · An example of linear interpolation that works perfectly.

Curve Fitting

Curve fitting provides an intuitive scheme for approximating a point-to-point graph, or filling in a graph containing one or more gaps, to make it look like a continuous curve. Figure 1-15 shows an approximate graph of the price of hypothetical stock Y, based on points determined at intervals of 1/2 hour as generated by curve fitting. Here, the dashed black line shows the moment-to-moment

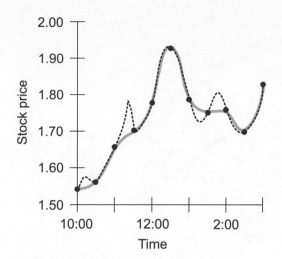

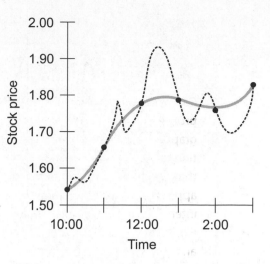

FIGURE 1-15 · Approximation of hypothetical stock price as a continuous function of time, making use of curve fitting. The solid gray curve represents the approximation; the black dashed curve represents the actual, moment-to-moment stock price as a function of time.

FIGURE 1-16 · An example of curve fitting in which insufficient data samples exist, causing significant errors. The solid gray curve represents the approximation; the black dashed curve represents the actual, moment-to-moment stock price as a function of time.

stock price, and the solid gray line shows the fitted curve based on half-hour values. The fitted curve does not precisely represent the actual stock price at every instant, but it comes close most of the time.

As we determine and plot the values at increasingly frequent intervals, we find that curve fitting gives us better and better results. When we determine and plot the values infrequently, curve fitting can introduce large errors. Figure 1-16 shows an example, where the price for stock Y shows up only at hourly intervals. Curve fitting obviously doesn't do a good job for us here!

Extrapolation

The term *extrapolate* means "to put outside of." When a function has a continuous-curve graph where time acts as the independent variable, *extrapolation* means essentially the same thing as *short-term forecasting*. Figure 1-17 shows two examples.

In Fig. 1-17A, we plot the price of the hypothetical stock X as a function of time until 2:00 p.m., and then we attempt to forecast its price for an hour into the future, based on its past performance. In this case, we use *linear extrapolation*, the simplest technique. We project the curve ahead as a straight line. Compare this graph with the solid curve in Fig. 1-6. In this case, linear extrapolation works fairly well.

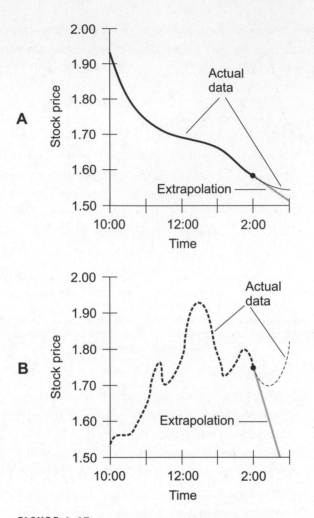

FIGURE 1-17 · Examples of linear extrapolation. The solid gray lines represent the forecasts; the black dashed curves represent the actual data. At A, we see a fairly good prediction. At B, we see an inaccurate linear extrapolation.

Figure 1-17B shows the price of the hypothetical stock Y plotted until 2:00 p.m., and then we employ linear extrapolation in an attempt to predict its behavior for the next hour. As you can see by comparing this graph with the dashed curve in Fig. 1-6, linear extrapolation does not work in this scenario!

Some graphs lend themselves well to extrapolation, while other graphs do not. In general, as a curve becomes more complicated, extrapolation becomes subject to more error. Also, as the extent (or distance) of the extrapolation increases for a given curve, the accuracy decreases, so you should expect the results to vary more and more from actual outcomes.

TIP   *If you want to take advantage of linear extrapolation, you can (and should) use computer software written expressly for that purpose. Machines can "see" subtle characteristics of functions that humans overlook.*

Trends

We call a function *nonincreasing* if the value of the dependent variable never grows any larger (or more positive) as the value of the independent variable increases. If the dependent variable in a function never gets any smaller (or more negative) as the value of the independent variable increases, we call that function *nondecreasing*.

The dashed curve in Fig. 1-18 shows the behavior of a hypothetical stock Q, whose price never rises throughout the period under consideration. This function is nonincreasing. The solid curve portrays the behavior of a hypothetical stock R, whose price never falls throughout the period. This function is nondecreasing.

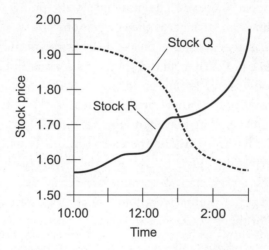

FIGURE 1-18 • The price of stock Q is nonincreasing versus time, and the price of stock R is nondecreasing versus time.

Don't Get Confused!

Sometimes, technical people use the terms *trending downward* and *trending upward* to describe graphs. These terms are subjective; different people might interpret them differently. Everyone would agree that stock Q in Fig. 1-18 trends downward while stock R trends upward. But we'll sometimes encounter a stock that rises and falls several times during a period, so we can't define it as either trending upward or trending downward.

Correlation

Specialized graphs called *scatter plots* can show the extent of *correlation* between the values of two variables when we obtain those values from a finite number of experimental samples.

If, as the value of one variable generally increases, the value of the other variable also generally increases, we consider the correlation positive. If the opposite holds true—the value of one variable generally increases as the other generally decreases—we consider the correlation negative. If the points appear randomly scattered all over the coordinate grid, then we consider the correlation as equal to 0 (nonexistent).

Figure 1-19 shows five examples of scatter plots. At A, no correlation exists. At B and C, we see positive correlation. At D and E, we see negative correlation. When correlation exists, the points tend to "bunch up" along a well-defined path. In these examples the paths constitute straight lines, but in some situations they can appear as curves. The more nearly the points in a scatter plot lie along a straight line, the stronger is the correlation.

Statisticians quantify correlation on a scale from a minimum of –1, through 0, up to a maximum of +1. When all the points in a scatter plot lie along a straight line that ramps downward as you go to the right, indicating that one variable decreases uniformly as the other variable increases, the correlation equals –1. When all the points lie along a straight line that ramps upward as you go to the right, indicating that one variable increases uniformly as the other variable increases, the correlation equals +1. None of the graphs in Fig. 1-19 show either of these extremes. Statisticians determine the actual value of the correlation factor for a given set of points according to a complicated algorithm that surpasses the scope of this course.

PROBLEM 1-8

Suppose, as the value of the independent variable in a function changes, the value of the dependent variable does not change. In this case, we have a *constant function*. Is its graph nonincreasing or nondecreasing?

SOLUTION

According to our definitions, the graph of a constant function is both nonincreasing and nondecreasing. Its value never increases, and it never decreases.

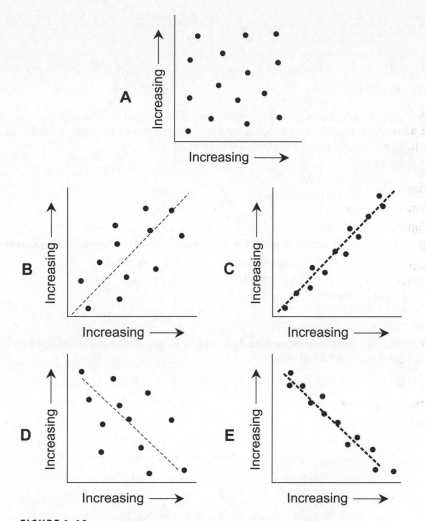

FIGURE 1-19 • Scatter plots showing correlation of 0 (at A), weak positive correlation (at B), strong positive correlation (at C), weak negative correlation (at D), and strong negative correlation (at E).

QUIZ

Refer to the text in this chapter if necessary. A good score is 8 correct. Answers are in the back of the book.

1. Suppose that we download a large file from the Internet. We plot the data transfer rate, in megabits per second, as a function of the time in seconds. In this situation, we would most likely consider time as
   A. the range.
   B. a nonincreasing value.
   C. the independent variable.
   D. a subset of the data.

2. We know for certain that we have a rational number when we can express it as
   A. a ratio of one whole number to another whole number.
   B. an endless repeating decimal.
   C. an even integer.
   D. All of the above

3. Figure 1-20 portrays three sets A, B, and C, along with six specific points P, Q, R, S, T, and U. Point R belongs to
   A. $(A \cap B) \cap C$.
   B. $A \cap B$.
   C. $A \cup B$.
   D. $A \cap C$.

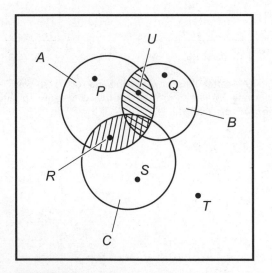

FIGURE 1-20 · Illustration for Quiz Questions 3, 4, and 5.

4. In the situation of Fig. 1-20, point *U* belongs to
    A. $(A \cap B) \cap C$.
    B. $A \cup B$.
    C. $B \cap C$.
    D. $A \cap C$.

5. In the situation of Fig. 1-20, one of the points does not belong to $(A \cup B) \cup C$. Which one?
    A. Point *P*
    B. Point *R*
    C. Point *T*
    D. Point *U*

6. In a vertical bar graph, we portray
    A. the independent variable on the vertical axis and the dependent variable on the horizontal axis.
    B. the independent variable on the horizontal axis and the dependent variable on the vertical axis.
    C. the independent variable as bar widths and the dependent variable as bar heights.
    D. the independent variable as bar heights and the dependent variable as bar widths.

7. We might reasonably suppose that the number of snowy days per year in a 10,000-square mile region correlates *negatively* with
    A. the number of days in the region during which the temperature remains above freezing.
    B. the number of storms per year in the region.
    C. the number of polar bears that live in the region.
    D. All of the above

8. Suppose that we attach a radio transmitter to a timber wolf and plot its location on a map at 15-minute intervals over a period of days. We might reasonably attempt to predict the wolf's path for an hour or two into the future using
    A. linear extrapolation.
    B. curve fitting.
    C. linear interpolation.
    D. a Venn diagram.

9. Which of the following statements holds true in general?
    A. All relations constitute functions.
    B. All functions constitute relations.
    C. The values in a bar graph always add up to 100%.
    D. Zero correlation is indicated by points that lie on a straight line in a scatter plot.

10. **Which of the following statements *always* holds true for a nondecreasing function?**

   A. The value of the dependent variable constantly increases or remains constant as the value of the independent variable increases.
   B. The value of the dependent variable constantly decreases or remains constant as the value of the independent variable increases.
   C. The value of the independent variable does not change.
   D. The value of the dependent variable does not change.

chapter **2**

Learning the Jargon

Statisticians analyze observable or measurable information, commonly known as *data*. We usually obtain statistical data by looking at the "real world," although we can also generate statistics in "artificial worlds" called *models*.

CHAPTER OBJECTIVES

In this chapter, you will

- Compare discrete and continuous variables
- Work with populations and samples
- Learn how to define and portray distributions
- Truncate and round off numerical values
- Determine mean, median, mode, and standard deviation

Experiments and Variables

If we want to understand a technical discipline, we must know the jargon. Let's define a few of the most widely used statistics terms.

Experiment

When we do a statistical *experiment*, we collect data with the intent of learning or discovering something. For example, we might conduct an experiment to determine the most popular channels for frequency-modulation (FM) radio broadcast stations having transmitters located in American towns of fewer than 10,000 population. We might conduct an experiment to determine the lowest barometric pressures in the centers of all Atlantic hurricanes that take place during a 10-year period.

Some experiments require specialized or complicated instruments to measure quantities. If we conduct an experiment to figure out the average test scores of high-school seniors in Wyoming who take a certain standardized test at the end of this school year, we need only time, energy, and willingness to collect the data. However, the measurement of the minimum pressure inside the eye of a hurricane requires sophisticated hardware in addition to time, energy, patience, and courage.

Variable (in General)

In mathematics, we define a *variable*, also called an *unknown*, as a quantity whose value we can figure out according to certain rules. We express mathematical variables by writing letters of the alphabet, usually in lowercase italics. For example, in the expression

$$x + y + z = 5$$

the letters x, y, and z represent variables. In this case, the variables represent numerical values (as opposed to, say, colors, or logic states).

Statistical variables resemble mathematical variables, but an important difference exists. In statistics, we always associate a variable with at least one experiment. In mathematics, we can work entirely in the theoretical realm, without doing any experiments whatsoever.

Discrete Variable

We define a *discrete variable* as a statistical variable that can attain only certain well-defined values. Discrete variables resemble the channels of a television set or digital broadcast receiver. We can easily express the value of a discrete variable, sometimes (but not always) with mathematical exactitude.

When a radio-station announcer says "You're listening to 97.1 FM," she means that the assigned channel center lies at a frequency of 97.1 *megahertz* (MHz), where 1 MHz represents 1,000,000 (10^6) *cycles per second* (or *hertz*, symbolized Hz). The assigned frequency constitutes an exact value, even though the broadcast engineer can at best make sure that the transmitter output frequency lies close to 97.1 MHz. In the United States, assigned channels in the FM broadcast band differ by an *increment* (minimum difference) of 0.2 MHz. The next lower channel from 97.1 MHz appears at 96.9 MHz, and the next higher channel appears at 97.3 MHz. We'll find no legally assigned channels in between those two frequencies. The lowest channel appears at 88.1 MHz and the highest channel appears at 107.9 MHz. Figure 2-1 illustrates this situation.

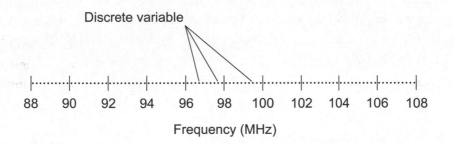

FIGURE 2-1 • The channels in the FM broadcast band constitute values of a discrete variable.

Other examples of discrete variables that can attain only specific, exact theoretical values include the following:

- The number of people voting for each candidate in a political election
- The scores of students who take a standardized test (expressed as a percentage of correct answers)
- The number of car drivers caught "speeding" every day in a certain town
- The particular face of a cubical die that turns up after we toss it once

Continuous Variable

A *continuous variable* can attain *infinitely many* values over a certain span or range. Instead of existing as specific values with an increment between any two, a continuous variable can change value to an arbitrarily small extent.

Continuous variables resemble the radio frequencies to which you can set an old-fashioned analog FM broadcast receiver. In this type of receiver, you can

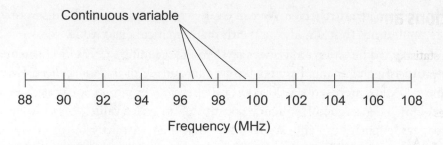

FIGURE 2-2 · The frequency to which we can set an analog FM broadcast receiver constitutes a continuous variable.

continuously adjust the frequency, say from 88 to 108 MHz (Fig. 2-2). If you move the tuning dial a little, you can make the received radio-frequency change by something less than 0.2 MHz, the separation between adjacent assigned transmitter channels. No limit exists as to how small the increment can get. If you have a light enough touch, you can adjust the received radio frequency up or down by an increment as small as 0.02 MHz, or 0.01 MHz, or even 0.0005 MHz. Other examples of continuous variables include:

- Temperature
- Barometric pressure
- Brightness of a light source
- Speed of a moving object
- Distance between two moving objects
- Time of day
- Sound-intensity levels

? Still Struggling

We can never exactly determine the *actual* values for continuous variables in the physical realm, even if we know their *theoretical* values with mathematical precision. This problem occurs whenever we use hardware to measure "real-world" parameters such as time, temperature, barometric pressure, and light-source brightness that can vary continuously. We'll always encounter some instrument or observation error in experimental sciences such as physics or electronics.

Populations and Samples

In statistics, the term *population* refers to a particular set of items, objects, phenomena, or people that we want to analyze. These items, also called *elements*, can be actual subjects such as people or animals, but they can also be numbers or definable quantities expressed in physical units. Examples of populations include the following:

- Assigned radio frequencies (in megahertz) of all FM broadcast transmitters in the United States
- Temperature readings (in degrees Celsius) at hourly intervals last Wednesday at various locations around the city of New York
- Minimum barometric-pressure levels (in millibars) at the centers of all the hurricanes in recorded history
- Brightness levels (in candela) of all the light bulbs in offices in Minneapolis, Minnesota
- Sound-intensity levels (in decibels relative to the threshold of hearing) of all the electric vacuum cleaners in the world

Sample, Event, and Census

A *sample* of a population constitutes a subset of that population. A sample can exist as a set containing only one value, reading, or measurement, singled out from a population. A sample might constitute a subset of a population that we can identify according to certain characteristics. The physical unit (if any) that defines a sample always matches the physical unit that defines the main, or parent, population. We call a single element of a sample an *event*. Examples of samples include:

- Assigned radio frequencies of FM broadcast stations whose transmitters are located in the state of Ohio
- Temperature readings at 1:00 p.m. local time last Wednesday at various locations around the city of New York
- Minimum barometric-pressure levels at the centers of Atlantic hurricanes during the decade from 1991 through 2000
- Brightness levels of compact fluorescent bulbs in offices in Minneapolis, Minnesota
- Sound-intensity levels of the electric vacuum cleaners used in all the households in Rochester, Minnesota

When a sample consists of the whole population, then we call it a *census*. When a sample consists of a subset of a population whose elements are chosen in a way that's as close to "random" as we can manage, then we call it a *random sample*.

Random Variable

A *random variable* is a discrete or continuous variable whose value we *cannot predict* in any given instance. In most circumstances, we'll define a random variable within a certain range of values, such as 1 through 6 for a tossed gambling die, or from 88 to 108 MHz in the case of an FM broadcast channel.

In a given scenario, we can often claim that some values of a random variable will *more likely* turn up than other values. In the case of a thrown die, assuming the die is not "weighted" to favor certain values over others, all of the values 1 through 6 are equally likely to turn up. When we consider the FM broadcast channels of public radio stations, we might find it tempting to suppose that certain types of transmissions occur more often at the lower radio-frequency range than at the higher range.

TIP *Have you ever noticed that a greater concentration of public radio stations appears in the 4-MHz-wide sample from 88 to 92 MHz than in, say, the equally wide sample from 100 to 104 MHz? It seems that way to me, but I haven't traveled around the country and spent the necessary time to confirm my hypothesis.*

? Still Struggling

In order for a variable to qualify as random, you must find it *impossible* to predict its value in any single instance. If you contemplate throwing a die (even a "weighted" die) only one time, you can't predict with absolute certainty how it will turn up. For example, if you contemplate throwing a dart only once at a map of the United States while wearing a blindfold, you can't know, in advance, the lowest assigned FM broadcast channel in the town nearest the point where the dart will hit.

Frequency

The *frequency* of a particular outcome (result) of an event is the number of times that outcome occurs within a specific sample of a population. Don't confuse statistical frequency with radio broadcast or computer processor frequency! In statistics, the term "frequency" means "often-ness." Two species of statistical frequency exist: *absolute frequency* and *relative frequency*.

Suppose you toss a die 6000 times. If the die is not "weighted," you should expect that the die will turn up showing one dot approximately 1000 times, two dots approximately 1000 times, and so on, up to six dots approximately 1000 times. The absolute frequency in such an experiment equals approximately 1000 for each face of the die. The relative frequency for each face equals approximately 1 in 6 or 16.67%.

Parameter

A *parameter* of a population constitutes a specific, well-defined characteristic of that population. We might want to know such parameters as the following, concerning populations mentioned earlier in this chapter:

- The most popular assigned FM broadcast frequency in the United States
- The highest temperature reading in the city of New York as determined at hourly intervals last Wednesday
- The average minimum barometric-pressure level or measurement at the centers of all the hurricanes in recorded history
- The lowest brightness level found in all the light bulbs in offices in Minneapolis, Minnesota
- The highest sound-intensity level found in all the electric vacuum cleaners used in the world

Statistic

We call a specific characteristic of a sample a *statistic* of that sample. We might want to know such statistics as:

- The most popular assigned frequency for FM broadcast stations in Ohio
- The highest temperature reading at 1:00 p.m. local time last Wednesday in New York
- The average minimum barometric-pressure level or measurement at the centers of Atlantic hurricanes during the decade from 1991 through 2000
- The lowest brightness level found in all the compact fluorescent bulbs in offices in Minneapolis, Minnesota
- The highest sound-intensity level found in electric vacuum cleaners used in households in Rochester, Minnesota

Distributions

We define a *statistical distribution* as a description of the set of possible values that a random variable can attain. We can define a distribution by noting the absolute or relative frequency. We can illustrate a distribution as a table or a graph, or both.

Discrete versus Continuous

Table 2-1 shows the results of a single, hypothetical experiment in which we toss an "unweighted" gambling die 6000 times. Figure 2-3 is a vertical bar graph of the data in Table 2-1. Both the table and the graph portray a distribution that

TABLE 2-1	Results of a single, hypothetical experiment in which we toss an "unweighted" die 6000 times.
Face of Die	**Number of Times Face Turns Up**
1	968
2	1027
3	1018
4	996
5	1007
6	984

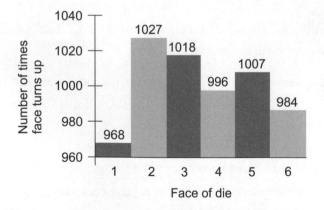

FIGURE 2-3 · Results of a hypothetical experiment in which we toss an "unweighted" die 6000 times.

TABLE 2-2	Number of days on which measurable rain occurs in a specific year, in five hypothetical towns.

Town Name	Number of Days in Year with Measurable Precipitation
Happyville	108
Joytown	86
Wonderdale	198
Sunnywater	259
Rainy Glen	18

describes the behavior of the die. If we repeat the experiment, the results will differ, of course. However, if we carry out a large number of these experiments, the relative frequency of each number turning up will approach 1 in 6, or approximately 16.67%.

Table 2-2 shows the number of days during the course of a 365-day year in which measurable precipitation occurs within the city limits of five hypothetical towns. Figure 2-4 is a horizontal bar graph showing the same data as Table 2-2. Again, both the table and the graph portray statistical distributions.

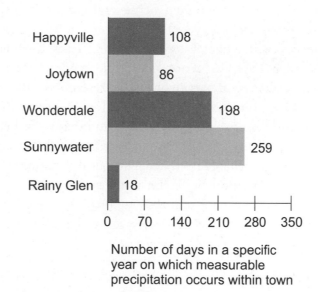

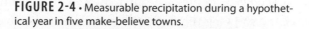
Number of days in a specific year on which measurable precipitation occurs within town

FIGURE 2-4 · Measurable precipitation during a hypothetical year in five make-believe towns.

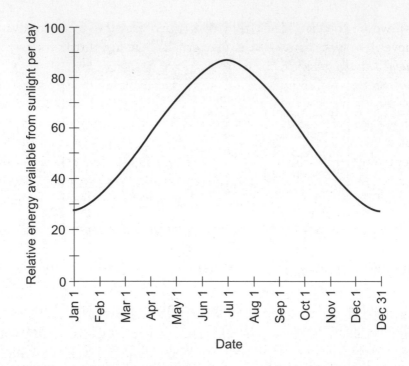

FIGURE 2-5 · Relative energy available per day from sunlight at a hypothetical location.

If we carry out the same experiment for several years in a row, we should expect the results to differ from year to year. Over a period of many years, the relative frequencies will converge toward certain values, although long-term climate change might have effects not predictable in our lifetimes.

Both of the preceding examples involve discrete variables. When we want to illustrate a distribution for a continuous variable (or a discrete variable that can attain a gigantic number of values), we'll often find that we're better off using a graph than a table. Figure 2-5 illustrates a distribution that denotes the relative amount of energy available from sunlight, per day during the course of a calendar year, at a hypothetical city in the northern hemisphere. We could construct a table showing this data, and that table would theoretically show us more detail than the graph of Fig. 2-5. However, that table would turn out messy indeed!

Frequency Distribution

In both of the foregoing examples (the first showing the results of 6000 die tosses and the second showing the days with precipitation in five hypothetical towns), we portray the scenarios with frequency as the dependent variable. Whenever frequency acts as the dependent variable in a distribution, we have a *frequency distribution*.

Suppose that we complicate the situation involving dice. Instead of one person tossing one die 6000 times, imagine five people tossing five different dice, and each person tosses the same die 6000 times. Suppose that the dice are colored red, orange, yellow, green, and blue, and are manufactured by five different companies, called Corp. A, Corp. B, Corp. C, Corp. D, and Corp. E, respectively. Further suppose that four of the dice are "weighted" and one is "unweighted." We therefore have a total of 30,000 die tosses to tabulate or graph.

Ungrouped Frequency Distribution

When we want to tabulate the die toss results as a frequency distribution, we can combine all the tosses and show the total frequency for each die face 1 through 6. Table 2-3 shows a hypothetical example of this result, called an *ungrouped frequency distribution*. We don't care about the weighting characteristics of each individual die, but only about the potential biasing of the entire set. It appears that, for this particular set of die, some bias exists in favor of faces 4 and 6, some bias exists against faces 1 and 3, and little or no bias exists either for or against faces 2 and 5.

TABLE 2-3 An ungrouped frequency distribution showing the results of a single, hypothetical experiment in which we toss five different dice, some "weighted" and some not "weighted," 6000 times.

Face of Die	Toss Results for All Dice
1	4857
2	4999
3	4626
4	5362
5	4947
6	5209

Grouped Frequency Distribution

If we want to render a distribution in more detail, we can tabulate the frequency for each face 1 through 6 separately for each die. Table 2-4 shows a hypothetical product of such an effort, which we call a *grouped frequency distribution*. We arrange the results according to the die manufacturer and color. From this distribution, we can see that some of the dice are heavily "weighted." Only the green die, manufactured by Corp. D, seems to lack bias. If you're astute, you'll strongly suspect that the green die here is the same die, with results gathered from the same experiment, as the one portrayed in Table 2-1 and Fig. 2-3.

TABLE 2-4   A grouped frequency distribution showing the results of a single, hypothetical experiment in which we toss five different dice manufactured by different companies, some dice "weighted" and some not "weighted," 6000 times.

| Face of Die | Toss Results by Manufacturer | | | | |
	Red Corp. A	Orange Corp. B	Yellow Corp. C	Green Corp. D	Blue Corp. E
1	625	1195	1689	968	380
2	903	1096	1705	1027	268
3	1300	890	1010	1018	408
4	1752	787	540	996	1287
5	577	1076	688	1007	1599
6	843	956	368	984	2058

PROBLEM 2-1

Suppose that you add up all the numbers in each vertical column of Table 2-4. What should you expect, and why? What should you expect if you repeat the experiment many times?

✔ SOLUTION

Each column should add up to 6000, representing the number of tosses for each die (red, orange, yellow, green, or blue) in the experiment. If you repeat the experiment many times, the sums of the numbers in each column should equal 6000 every time.

? Still Struggling

If the sum of the numbers in any of the columns in Table 2-4 does not equal 6000, then you performed the experiment in a faulty fashion, or else you made an error in the compilation of the table.

PROBLEM **2-2**

Suppose that you add up all the numbers in each horizontal row of Table 2-4. What should you expect, and why? What should you expect if you repeat the experiment many times?

SOLUTION

The sums of the numbers in the rows will vary, depending on the bias of the set of dice considered as a whole. If, taken all together, the dice show any bias, and if you conduct the experiment numerous times, the sums of the numbers should turn out consistently lower for some rows than for other rows.

PROBLEM **2-3**

Each small rectangle in a table, representing the intersection of one row with one column, is called a *cell* of the table. What do the individual numbers in the cells of Table 2-4 represent?

SOLUTION

Each individual number represents an absolute frequency—the number of times a particular face of a particular die came up during the course of the experiment.

More Definitions

Statistics has no shortage of jargon! Following are some more definitions that you should learn in order to remain comfortable in your encounters with statisticians.

Truncation

The process of *truncation* allows us to approximate the values of numbers portrayed as decimal expansions. Truncation involves the deletion of all the numerals to the right of a certain point in the decimal part of an expression. Some electronic calculators use truncation to fit numbers within their displays. For example, we can use truncation to "whittle down" the number 3.830175692803 in steps as follows:

3.830175692803

3.83017569280

3.8301756928

3.830175692

3.83017569

3.8301756

3.830175

3.83017

3.8301

3.830

3.83

3.8

3

Rounding

Rounding is the preferred method of approximating numbers when we find them denoted as decimal expansions. In this process:

- When we delete a given digit (call it r) at the right-hand extreme of an expression and $0 \le r \le 4$, we don't change the digit q to its left (which becomes the new r after we've deleted the old r).

- When we delete a given digit (call it r) at the right-hand extreme of an expression and $5 \le r \le 9$, we increase the digit q to its left by 1 (we "round up").

Most electronic calculators employ rounding rather than truncation in order to approximate numerical values to a fixed number of significant digits. In the rounding process, the number 3.830175692803 "erodes away" in steps as follows:

$$3.830175692803$$
$$3.83017569280$$
$$3.8301756928$$
$$3.830175693$$
$$3.83017569$$
$$3.8301757$$
$$3.830176$$
$$3.83018$$
$$3.8302$$
$$3.830$$
$$3.83$$
$$3.8$$
$$4$$

Cumulative Absolute Frequency

When we tabulate data, we'll often want to portray the absolute frequencies in one or more columns. Table 2-5 shows an example. Here, we see the results of the tosses of the blue die in the experiment we looked at a few moments ago. The first column shows the number on the die face. The second column shows

TABLE 2-5 Results of an experiment in which we toss a "weighted" die 6000 times, showing absolute frequencies and cumulative absolute frequencies.

Face of Die	Absolute Frequency	Cumulative Absolute Frequency
1	380	380
2	268	648
3	408	1056
4	1287	2343
5	1599	3942
6	2058	6000

the absolute frequency for each face, or the number of times each face turned up during the experiment. The third column shows the *cumulative absolute frequency*: the sum of all the absolute frequency values in table cells at or above the given position.

? Still Struggling

The cumulative absolute frequency numbers in a table always ascend (increase) as we go down the column. The total cumulative absolute frequency should equal the sum of all the individual absolute frequency numbers. In this instance, that sum comes out to 6000, the number of times that we tossed the blue die.

Cumulative Relative Frequency

We can add up relative frequency values going down the columns of a table, in exactly the same way as we add up the absolute frequency values. When we do this, the resulting values (usually expressed as percentages) tell us the *cumulative relative frequency*.

Table 2-6 offers a detailed analysis of what happened with the blue die in the above-described experiment. The first, second, and fourth columns in Table 2-6

TABLE 2-6	Results of an experiment in which we toss a "weighted" die 6000 times, showing absolute frequencies, relative frequencies, cumulative absolute frequencies, and cumulative relative frequencies.			
Face of Die	Absolute Frequency	Relative Frequency (%)	Cumulative Absolute Frequency	Cumulative Relative Frequency (%)
1	380	6.33	380	6.33
2	268	4.47	648	10.80
3	408	6.80	1056	17.60
4	1287	21.45	2343	39.05
5	1599	26.65	3942	65.70
6	2058	34.30	6000	100.00

coincide with the first, second, and third columns in Table 2-5. The third column in Table 2-6 shows the percentage represented by each absolute frequency number. We get these percentages by dividing each number in the second column by 6000, the total number of tosses. The fifth column shows the cumulative relative frequency, which equals the sum of all the relative frequency values in table cells at or above the given position.

TIP   *The cumulative relative frequency percentages in a table, like the cumulative absolute frequency numbers, always ascend as we go down the column. The total cumulative relative frequency should equal 100%. In this sense, the cumulative relative frequency column in a table can serve as a* **checksum,** *helping to ensure that we've tabulated the entries correctly.*

Mean

The *mean* for a discrete variable in a distribution is the mathematical average of all the values. If we consider a variable over the entire population, we call the average value the *population mean*. If we consider the variable over a particular sample of a population, we call the average value the *sample mean*. We can only have one population mean for a population, but many different sample means can exist. Statisticians denote the mean as a lowercase Greek letter *mu* in italics (μ). Alternatively, some people denote the mean by writing an italicized lowercase English letter, usually *x*, with a bar (vinculum) over it.

Table 2-7 shows the results of a hypothetical 10-question test, given to a class of 100 students. As you can see, every possible score is accounted for. Some people answered all 10 questions correctly; a few people didn't get a single answer right. In order to determine the mean score for the whole class on this test—the population mean, called μ_p—we must add up the scores of each and every student, and then divide that sum by 100. First, let's sum up the products of the numbers in the first and second columns. This process yields 100 times the population mean, as follows:

$$(10 \times 5) + (9 \times 6) + (8 \times 19) + (7 \times 17) + (6 \times 18) + (5 \times 11)$$
$$+ (4 \times 6) + (3 \times 4) + (2 \times 4) + (1 \times 7) + (0 \times 3)$$
$$= 50 + 54 + 152 + 119 + 108 + 55 + 24 + 12 + 8 + 7 + 0$$
$$= 589$$

TABLE 2-7   Scores on a hypothetical 10-question test taken by 100 students.

Test Score	Absolute Frequency	Letter Grade
10	5	A
9	6	A
8	19	B
7	17	B
6	18	C
5	11	C
4	6	D
3	4	D
2	4	F
1	7	F
0	3	F

Dividing this value by 100, the total number of test scores (one for each student who turns in a paper), we obtain

$$\mu_p = 589/100$$
$$= 5.89$$

The teacher in this class has assigned letter grades to each score. Students who scored 9 or 10 correct received grades of A; students who got scores of 7 or 8 received grades of B; those who got scores of 5 or 6 got grades of C; those who got scores of 3 or 4 got grades of D; those who got fewer than 3 correct answers received grades of F. The assignment of grades, informally known as the "curve," depends on the temperament of the teacher.

PROBLEM 2-4

What are the sample means for each grade in the test whose results appear in Table 2-7? Use rounding to determine the answers to two decimal places.

✔ SOLUTION

Let's call the sample means μ_{sa} for the grade of A, μ_{sb} for the grade of B, and so on down to μ_{sf} for the grade of F.

To calculate μ_{sa}, note that five students received scores of 10, while six students got scores of 9, both good enough for an A. This gives us a total of 5 + 6, or 11, students getting the grade of A. Therefore

$$\mu_{sa} = [(5 \times 10) + (6 \times 9)]/11$$
$$= (50 + 54)/11$$
$$= 104/11$$
$$= 9.45$$

To find μ_{sb}, observe that 19 students scored 8, and 17 students scored 7. Thus, 19 + 17, or 36, students received grades of B. Calculating, we obtain

$$\mu_{sb} = [(19 \times 8) + (17 \times 7)]/36$$
$$= (152 + 119)/36$$
$$= 271/36$$
$$= 7.53$$

To determine μ_{sc}, we can check the table to see that 18 students scored 6, while 11 students scored 5. Therefore, 18 + 11, or 29, students did well enough for a C. Grinding out the numbers yields

$$\mu_{sc} = [(18 \times 6) + (11 \times 5)]/29$$
$$= (108 + 55)/29$$
$$= 163/29$$
$$= 5.62$$

To calculate μ_{sd}, we note that six students scored 4, while four students scored 3. This means that 6 + 4, or 10, students got grades of D, so we have

$$\mu_{sd} = [(6 \times 4) + (4 \times 3)]/10$$
$$= (24 + 12)/10$$
$$= 36/10$$
$$= 3.60$$

Finally, we determine μ_{sf}. We see that four students got scores of 2, seven students got scores of 1, and three students got scores of 0. Thus, 4 + 7 + 3, or 14, students failed the test. It follows that

$$\mu_{sf} = [(4 \times 2) + (7 \times 1) + (3 \times 0)]/14$$
$$= (8 + 7 + 0)/14$$
$$= 15/14$$
$$= 1.07$$

Median

We define the *median* for the discrete variable in a distribution as the value such that the number of elements greater than or equal to it is the same as the number of elements less than or equal to it. In this context, the term *median* means "middle" or "midway."

Table 2-8 shows the results of the 10-question test described above, but instead of showing letter grades in the third column, we see the cumulative absolute frequencies. We begin the tally with the top scoring papers and proceed

TABLE 2-8 We can determine a median by tabulating cumulative absolute frequencies.

Test Score	Absolute Frequency	Cumulative Absolute Frequency
10	5	5
9	6	11
8	19	30
7	17	47
6 (partial)	3	50
6 (partial)	15	65
5	11	76
4	6	82
3	4	86
2	4	90
1	7	97
0	3	100

in order downward. (We could just as well do it the other way, starting with the lowest-scoring papers and proceeding upward.) When we tally the scores of all 100 individual papers this way so that they appear in order, the scores of the 50th and 51st papers—the two in the middle—equal 6 correct. Thus, the median is 6, because half the students scored 6 or above, and the other half scored 6 or below.

? Still Struggling

It's possible that in another group of 100 students taking this same test, the 50th paper would have a score of 6 while the 51st paper would have a score of 5. When two values "compete," we define the median as their average. In this case the median would fall midway between 5 and 6, so we'd call it 5.5.

Mode

We define the *mode* for a discrete variable as the value that occurs the most often. In the test whose results appear in Table 2-7, the score of 8 appeared more often than any other score. Nineteen students got that score. No other score had that many results. Therefore, in this situation, the mode equals 8.

What if another group of students took this test, and two different scores occurred equally often? For example, suppose that 16 students scored 8, and 16 students scored 6. In this case, two modes would exist: 6 and 8. We'd have a so-called *bimodal distribution*. We could even encounter a scenario with three or more modes (a *multimodal distribution*).

Finally, imagine that the class had only 99 students on a certain day (one of them stayed home sick), and nine students got each of the 11 possible scores (from 0 to 10 correct answers). In this distribution, no mode would exist at all.

Measures of Central Tendency

Statisticians call the mean, median, and mode *measures of central tendency*, because these parameters describe, in three different ways, a sort of "center of gravity" for the values in a data set.

Variance

We can describe the nature of a distribution in still another way: measure or express the extent to which the values appear "spread-out." A distribution of test scores such as those portrayed in Table 2-7 differs qualitatively from a distribution where the scores occur with almost equal frequency. The test results portrayed in Table 2-7 also differ qualitatively from a distribution where almost every student gets the same score.

In the scenario of Table 2-7, let's call the variable x. Let's call the individual students' scores x_1 through x_{100}. Suppose that we take the time to figure out the extent to which each individual student's score x_i (where i equals an integer between 1 and 100) differs from the mean score for the whole population (μ_p). This process gives us 100 "distances from the mean," d_1 through d_{100}, as follows:

$$d_1 = |x_1 - \mu_p|$$
$$d_2 = |x_2 - \mu_p|$$
$$\downarrow$$
$$d_{100} = |x_{100} - \mu_p|$$

? Still Struggling

Vertical lines on each side of an expression represent the *absolute value*, the extent to which that expression differs from 0. For any particular real-number expression r, we define

$$|r| = r \text{ if } r \geq 0$$

and

$$|r| = -r \text{ if } r < 0$$

The absolute-value function allows us to avoid the occurrence of negative numbers when they would cause confusion or ambiguity.

Let's square each of the above-defined "distances from the mean" to obtain the following set of numbers:

$$d_1^2 = (x_1 - \mu_p)^2$$
$$d_2^2 = (x_2 - \mu_p)^2$$
$$\downarrow$$
$$d_{100}^2 = (x_{100} - \mu_p)^2$$

We don't need absolute-value symbols in these expressions, because the squares of all real numbers turn out nonnegative.

Now let's average all the "squares of the distances from the mean," d_i^2. In other words, let's add them all up and then divide that sum by 100 (the total number of scores) to get the "average of the squares of the distances from the mean." We call the resulting quantity the *variance* of the variable x, written Var(x):

$$\text{Var}(x) = (1/100)\ (d_1^2 + d_2^2 + \cdots + d_{100}^2)$$
$$= (1/100)\ [(x_1 - \mu_p)^2 + (x_2 - \mu_p)^2 + \cdots + (x_{100} - \mu_p)^2]$$

If we have a set of n values whose population mean equals μ_p, then we can calculate the variance using the general formula

$$\text{Var}(x) = (1/n)\ [(x_1 - \mu_p)^2 + (x_2 - \mu_p)^2 + \cdots + (x_n - \mu_p)^2]$$

Standard Deviation

Standard deviation, like variance, expresses the extent to which values appear "spread-out" with respect to the mean. The standard deviation equals the square root of the variance. We can symbolize standard deviation by writing an italicized, lowercase Greek letter sigma (σ). In the scenario of the examination described earlier in this chapter, we have

$$\sigma = [(1/100)(d_1^2 + d_2^2 + \cdots + d_{100}^2)]^{1/2}$$
$$= \{(1/100)\ [(x_1 - \mu_p)^2 + (x_2 - \mu_p)^2 + \cdots + (x_{100} - \mu_p)^2]\}^{1/2}$$

The standard deviation of a set of n values whose population mean equals μ_p is given by the formula

$$\sigma = \{(1/n)\ [(x_1 - \mu_p)^2 + (x_2 - \mu_p)^2 + \cdots + (x_n - \mu_p)^2]\}^{1/2}$$

? Still Struggling

If the foregoing formulas baffle you, don't worry. You don't have to memorize them (although doing so won't harm you). You can remember the following verbal definitions:

- The variance equals the average of the squares of the "distances" of each value from the mean.
- The standard deviation equals the square root of the variance.

TIP *Because variance equals the square of standard deviation, you'll occasionally see variance symbolized as* σ^2. *Variance and standard deviation both constitute parameters known as* measures of dispersion. *In this context, the term "dispersion" means "spread-outedness."*

PROBLEM 2-5

Draw a vertical bar graph showing all the *absolute-frequency* data from Table 2-5, the results of a "weighted" die-tossing experiment. Portray each die face on the horizontal axis. Use light gray vertical bars to show the absolute frequency numbers. Use dark gray vertical bars to show the cumulative absolute frequency numbers.

SOLUTION

Figure 2-6 shows such a graph. To avoid excessive clutter, the numerical data doesn't appear at the tops of the bars.

PROBLEM 2-6

Draw a horizontal bar graph showing all the *relative-frequency* data from Table 2-6, another portrayal of the results of a "weighted" die-tossing experiment. Show each die face on the vertical axis. Use light gray horizontal bars to show the relative frequency percentages. Use dark gray horizontal bars to show the cumulative relative frequency percentages.

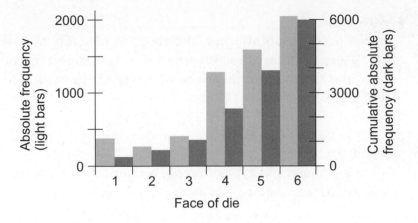

FIGURE 2-6 · Illustration for Problem 2-5.

![checkmark] **SOLUTION** _____

Figure 2-7 shows the graph. In the interest of neatness, the individual percentages do not appear at the ends of the bars.

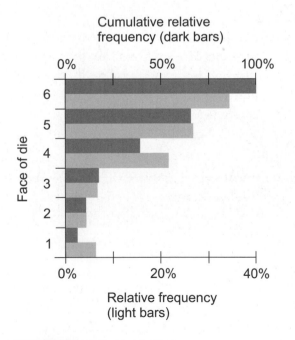

FIGURE 2-7 · Illustration for Problem 2-6.

PROBLEM 2-7

Draw a point-to-point graph showing the *absolute frequencies* of the 10-question test described by Table 2-7. Mark the population mean, the median, and the mode with distinctive vertical lines, and label them.

SOLUTION

Figure 2-8 shows this type of graph. Numerical data appears for the population mean, median, and mode.

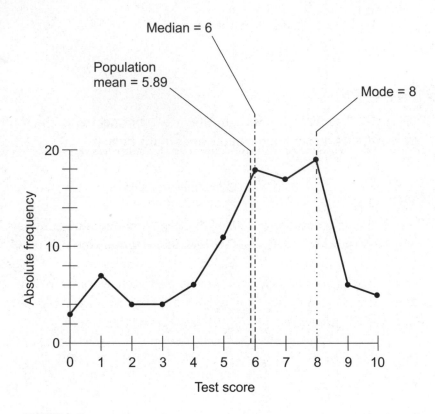

FIGURE 2-8 · Illustration for Problem 2-7.

PROBLEM 2-8

Calculate the variance, Var(x), for the 100 test scores tabulated in Table 2-7.

TABLE 2-9 "Distances" of each test score x_i from the population mean μ_p, the squares of these "distances," the products of the squares with the absolute frequencies f_i, and the sums of these products. This information applies to Solution 2-8.

Test Score	Abs. Freq. f_i	"Distance" from Mean: $x_i - \mu_p$	"Distance" Squared: $(x_i - \mu_p)^2$	Product: $f_i \times (x_i - \mu_p)^2$
10	5	+4.11	16.89	84.45
9	6	+3.11	9.67	58.02
8	19	+2.11	4.45	84.55
7	17	+1.11	1.23	20.91
6	18	+0.11	0.01	0.18
5	11	−0.89	0.79	8.69
4	6	−1.89	3.57	21.42
3	4	−2.89	8.35	33.40
2	4	−3.89	15.13	60.52
1	7	−4.89	23.91	167.37
0	3	−5.89	34.69	104.07
Sum of products				643.58

 SOLUTION

Recall that the population mean, μ_p, as determined earlier, equals 5.89. Table 2-9 shows the "distances" of each score from μ_p, the squares of these "distances," and the products of each of these squares with the absolute frequencies (the number of papers having each score from 0 to 10). At the bottom of the table, we sum all of these products. The resulting number, 643.58, equals 100 times the variance. Therefore

$$\text{Var}(x) = (1/100)(643.58)$$

$$= 6.4358$$

We can round this off to 6.44.

PROBLEM **2-9**

Calculate the standard deviation for the 100 test scores tabulated in Table 2-7.

SOLUTION

The standard deviation, σ, equals the square root of the variance. Approximating on the basis of the "un-rounded" value for the variance that we derived above, we get

$$\sigma = [\text{Var}(x)]^{1/2}$$

$$= 6.4358^{1/2}$$

$$= 2.5369$$

We can round this off to 2.54.

QUIZ

Refer to the text in this chapter if necessary. A good score is 8 correct. Answers are in the back of the book.

1. Suppose that a variable n can attain a value equal to any positive even whole number less than or equal to 500. We can call n
   A. an absolute variable.
   B. a discrete variable.
   C. a continuous variable.
   D. a limited variable.

2. Which of the following terms describes a value for a discrete variable that occurs the most frequently?
   A. The average
   B. The mode
   C. The median
   D. The variance

3. Imagine that 21 people take a test consisting of 20 questions, and each student gets a different score from all the other students. That is, one student scores 0, one student scores 1, one student scores 2, and so on. What's the mean score?
   A. We can't define a specific mean.
   B. 20
   C. The square root of 20
   D. 10

4. Which of the following parameters constitutes a measure of central tendency as defined in this chapter?
   A. The median
   B. The standard deviation
   C. The cumulative frequency
   D. The variance

5. Which of the following parameters constitutes a measure of dispersion as defined in this chapter?
   A. The mean
   B. The median
   C. The mode
   D. The variance

6. Imagine that 37 people take a test consisting of 36 questions, and each student gets a different score from all the other students. That is, one student scores 0, one student scores 1, one student scores 2, and so on. What's the median?

    A. 17
    B. 17.5
    C. 18
    D. We can't define it.

7. Imagine that 44 people take a 15-question test, and 22 people students score 8 correct while the other 22 students score 10 correct. What's the mode?

    A. Both 8 and 10
    B. 9
    C. The square root of 80
    D. We can't define it.

8. In the scenario of Question 7, what's the median?

    A. Both 8 and 10
    B. 9
    C. The square root of 80
    D. We can't define it.

9. In the scenario of Question 7, what's the variance, accurate to three decimal places?

    A. 2.000
    B. 1.414
    C. 1.000
    D. We can't define it.

10. In the scenario of Question 7, what's the standard deviation, accurate to three decimal places?

    A. 2.000
    B. 1.414
    C. 1.000
    D. We can't define it.

Basics of Probability

Probability quantifies the proportion of the time that specific things happen. We can also define *probability* as the science that allows us to determine the proportion of the time that specific things happen.

CHAPTER OBJECTIVES

In this chapter, you will

- Learn when probability makes sense, and when it doesn't
- Compare events and outcomes
- Contrast mathematical probability with empirical probability
- Calculate permutations and combinations
- Determine probability density functions
- Express probability in the form of a distribution

The Probability Fallacy

We say that a certain fact holds true because we've seen or deduced it. If we believe that something is happening or has happened but we can't prove it, we find it tempting to say that the event "is likely" or that it "probably happened."

Belief

When people formulate a theory, they often say that something "probably" took place in the distant past, or that something "might" exist somewhere, as-yet undiscovered, at this moment. Have you ever heard that there's a "good chance" that extraterrestrial life exists? Such a statement has no meaning in a pure mathematical sense. Either extraterrestrial life exists, or it does not.

If you say "I suspect that the universe began with an explosion," you state the fact that you believe it, not the fact that it's true or that it's "probably" true. If you say "The universe began with an explosion," you make a logically sound statement, but you state a theory, not a proven fact. If you say "The universe probably started with an explosion," you in effect suggest that multiple pasts existed, and our universe had an explosive origin in most of them. A statement like that constitutes an example of the *probability fallacy* (PF). We commit this fallacy when we inject probability into a discussion or scenario where it has no legitimate role.

Whatever is, is. Whatever isn't, isn't. Whatever was, was. Whatever wasn't, wasn't. Either the universe started with an explosion, or else it didn't. Either life exists on some other world, or else it doesn't. Single-event situations have no "in-between" states unless you spin theories involving multiple or parallel realities.

If we say that the "probability" of life existing elsewhere in the cosmos equals 20%, we say, in effect, "Out of n observed universes, where n represents some whole number, $0.2n$ universes contain extraterrestrial life." That proposition means nothing to those of us who have seen only one universe! If we say "It's probably raining in London right now," we insinuate that things can happen in more than one way as we speak. A phone call to London can resolve that dichotomy.

Fuzzy Truth

In mathematical logic, we will find systems that deal with so-called *fuzzy truth*, in which events can "sort of happen" or things can "hold true more or less." These theories involve degrees of truth that span a range, or *continuum*, over which we can assign probability values to specific occurrences. *Quantum mechanics* offers an example, in which theoretical physicists describe the behavior of subatomic particles in terms of probabilities or uncertain values.

? Still Struggling

In many situations, we can define probability according to the results of observations, although we can also define it on the basis of theory alone. When we abuse the notion of probability, however, seemingly sound reasoning can lead to absurd conclusions. Industrial professionals or politicians do this sort of thing every day, especially when they want to get somebody (such as you) to do something that will cause somebody else (such as them) to make money or gain power. Keep your "probability fallacy radar" on when navigating through any zero-sum game!

TIP   *If you come across an instance where an author says that something "probably happened," "is probably true," "probably took place," or "is not likely to happen," then the author* believes *that something holds true, occurred, or won't occur. Only observation or experimentation can resolve the issue. For example, I can tell you straightaway that I'm* probably *going to make statements later in this book to which you should apply this clarification. Maybe I've already done so. If you want to find out for sure, you must go through this entire book and check every single sentence right now. I won't blame you if you decide not to bother.*

Key Definitions

Let's define some common terms that allow us to write and talk about probability in a meaningful, comprehensible way.

Event versus Outcome

An *event* comprises a single test or trial in an experiment, or in the course of multiple identical experiments. We define an *outcome* as the result of an event verified by observation. For example, if you toss a coin 100 times, you have

100 separate events; each event constitutes a single toss of the coin. Each outcome equals either "heads" or "tails." If you throw a pair of dice simultaneously 50 times, each act of throwing the pair creates an event, so you have 50 events. You can define each outcome in terms of the number of dots facing upward when the two dice land on a given toss.

Imagine that, as you toss a coin repeatedly, you assign "heads" a value of 1 and "tails" a value of 0. When the coin comes up "heads," you say that the outcome of that event equals 1. If it comes up "tails," you say that the outcome equals 0. Suppose that as you toss a pair of dice repeatedly, you define the outcome as the total number of black dots that face upward when the dice come to rest. If you throw a pair of dice and see four dots facing up on one die and three dots facing up on the other die, then the outcome of that event equals 7.

As you can imagine, the outcome of an event depends on the nature of the hardware and processes involved in the experiment. A magnetized coin will behave differently, if you hurl it at a magnetized surface, than an ordinary "fair" coin will behave when you toss it onto a table. A pair of "weighted" or "unfair" dice will produce different outcomes, for an identical set of events, than will a pair of "unweighted" or "fair" dice.

The outcome of an event also depends on how you define that event in the first place. You can say, as above, that the outcome equals 7 for a particular toss of two dice. Alternatively, you can talk about the more specific instance in which one die comes up showing 3 while the other die comes up showing 4. You can go into even further detail by citing the case in which the die on your left shows 3 and the die on your right shows 4.

Sample Space

We define a *sample space* as the set of all possible outcomes in an experiment, or in the course of multiple identical experiments. It's a theoretical concept until we start to observe the outcomes.

Even if only a few events take place, we can have a large sample space. For example, if we toss a coin four times, we have 16 possible outcomes. Table 3-1 lists these outcomes, where "heads" = 1 and "tails" = 0. (If the coin happens to land on its edge, then we disregard that result and toss it again.)

If we toss a pair of dice—say one red die and one blue die—one time, we have 36 possible outcomes in the sample space as shown in Table 3-2. We can depict the outcomes as numerical *ordered pairs*, with the face-number of the red die listed first and the face-number of the blue die listed second.

TABLE 3-1 The sample space for an experiment in which we toss a coin four times. We can observe 16 possible outcomes; "heads" = 1 and "tails" = 0.

Event 1	Event 2	Event 3	Event 4
0	0	0	0
0	0	0	1
0	0	1	0
0	0	1	1
0	1	0	0
0	1	0	1
0	1	1	0
0	1	1	1
1	0	0	0
1	0	0	1
1	0	1	0
1	0	1	1
1	1	0	0
1	1	0	1
1	1	1	0
1	1	1	1

TABLE 3-2 The sample space for an experiment consisting of a single event in which we toss a pair of dice (one red, one blue) once. We can observe 36 possible outcomes, shown as ordered pairs (red,blue).

Red →	1	2	3	4	5	6
Blue ↓						
1	(1,1)	(2,1)	(3,1)	(4,1)	(5,1)	(6,1)
2	(1,2)	(2,2)	(3,2)	(4,2)	(5,2)	(6,2)
3	(1,3)	(2,3)	(3,3)	(4,3)	(5,3)	(6,3)
4	(1,4)	(2,4)	(3,4)	(4,4)	(5,4)	(6,4)
5	(1,5)	(2,5)	(3,5)	(4,5)	(5,5)	(6,5)
6	(1,6)	(2,6)	(3,6)	(4,6)	(5,6)	(6,6)

Mathematical Probability

Consider a discrete variable x that can attain n possible values, all equally likely on the basis of past observation. Suppose that an outcome h results from exactly m different values of x, where $m \leq n$. We can determine the *mathematical probability* $p_{math}(h)$ that outcome h will result from any given value of x using the formula

$$p_{math}(h) = m/n$$

Expressed as a percentage, it's

$$p_{math\%}(h) = 100 \, m/n$$

If we toss an "unweighted" die once, each of the six faces is just as likely to turn up as each of the others. We can expect to see 1 exactly as often as we'll see 2, 3, 4, 5, or 6. In this case, we have six possible values, so $n = 6$. We calculate the mathematical probability of any particular face turning up by setting $m = 1$, thereby obtaining

$$p_{math}(h) = 1/6$$

To calculate the mathematical probability that either of any two different faces will turn up (say 3 or 5), we set $m = 2$, getting

$$p_{math}(h) = 2/6$$

$$= 1/3$$

If we want to know the mathematical probability that any one of the six faces will turn up, we set $m = 6$, so the formula gives us

$$p_{math}(h) = 6/6$$

$$= 1$$

The respective percentages $p_{math\%}(h)$ in these cases equal 16.67% (approximately), 33.33% (approximately), and 100% (exactly).

Don't Get Confused!

Mathematical probabilities can only exist within the range 0 to 1 (or 0% to 100%) inclusive. The following formulas describe this constraint:

$$0 \leq p_{math}(h) \leq 1$$

and

$$0\% \leq p_{math\%}(h) \leq 100\%$$

We can never have a mathematical probability of 2 or −45%, or −6 or 556%. An outcome can't happen less often than never, or more often than all the time!

Empirical Probability

In order to determine the likelihood that an event will have a certain outcome in real life, we must rely on the results of prior experiments. We define the probability of a particular outcome taking place, based on experience or observation, as *empirical probability*.

Imagine that someone says a die is "unweighted." How does she know it? If we want to use this die in some application, such as when we need an object that can help us to generate a string of *pseudorandom numbers* from the set {1, 2, 3, 4, 5, 6}, we can't take on faith the notion that the die is "fair"; we must test it. We can analyze the die in a lab and figure out where its center of gravity lies; we can measure the depths of the indentations for the dots on its faces. We can scan the die electronically, X-ray it, and submerge it in (or float it on) water. But to gain a real measure of faith that in fact we have an "unweighted" die, we must toss it thousands or millions of times and confirm that each face turns up, on the average, 1/6 of the time. The more tosses, the better! We must conduct an experiment—gather *empirical evidence*—supporting the contention that we have a "fair" die. Statisticians base empirical probability values on determinations of relative frequency, which we discussed in Chap. 2.

As with mathematical probability, limits exist as to the range that an empirical probability figure can attain. If h represents an outcome for a particular single event, and if we denote the empirical probability of h taking place as a result of that event by writing $p_{emp}(h)$, then

$$0 \leq p_{emp}(h) \leq 1$$

and

$$0\% \leq p_{emp\%}(h) \leq 100\%$$

PROBLEM 3-1

Suppose that a new cholesterol-lowering drug comes on the market. If the drug is to gain approval for public use, its manufacturer must prove it effective and must also demonstrate a lack of serious side effects. Experimenters conduct extensive tests on the drug. During the course of testing, 10,000 people, all of whom have been diagnosed with high cholesterol levels in their bloodstreams, receive controlled doses of the drug. Imagine that 7289 of these 10,000 people experience a significant drop in their cholesterol levels. Also suppose that 307 people experience adverse side effects. If you have high cholesterol and you decide to take this drug, what's the empirical probability that your cholesterol level will go down? What's the empirical probability that you will experience adverse side effects?

SOLUTION

Some readers will say that we can't satisfactorily answer this question because the experiment doesn't go far enough. Does 10,000 represent a large enough number of test subjects? What physiological factors affect the way the drug works? How about the blood types of the individual subjects? Their ethnicity? Their gender? Their blood pressure? Their dietary habits? The extent to which they exercise? What, exactly, constitutes "high cholesterol"? What constitutes a "significant drop" in cholesterol level? What constitutes an "adverse side effect"? What's the standard drug dose? How long must a person take the drug in order to know whether it works or not? How does high cholesterol respond to the "placebo effect," a phenomenon in which people sometimes experience reactions to empty pills based on what they think will happen with the real drug? For the sake of simplicity (and our sanity!), let's ignore all of these factors here, even though, in a true scientific experiment, we would have to consider them.

Based on the foregoing experimental data (as simplified by eliminating all the "nuisance questions"), we can calculate the relative frequency of effectiveness as

$$7289/10{,}000 = 0.7289$$

$$= 72.89\%$$

and the relative frequency of ill effects as

$$307/10{,}000 = 0.0307$$

$$= 3.07\%$$

Let's round these values off to 73% and 3%, respectively. They represent the empirical probabilities that *you* will derive benefit, or experience adverse effects, if *you* take this particular drug in the hope of lowering your blood cholesterol level. Of course, once you actually consume the drug, these probabilities will lose meaning. You'll eventually say either "The drug worked for me" or "The drug didn't work for me." Your doctor will either claim that you experienced bad side effects, or conclude that you didn't.

Real-World Empiricism

Scientists use empirical probability figures to make predictions. Such figures do not, however, work very well for scrutinizing the past or present. If you try to calculate the empirical probability of the existence of extraterrestrial life in our galaxy, you can play around with formulas based on expert opinions, but once you state a numerical figure, you commit the PF. If you say that the empirical probability that a hurricane of category 3 or stronger struck the U.S. mainland in 1992 equals $x\%$ (where $x < 100$) because at least one hurricane of that intensity hit the U.S. mainland in x of the years in the 20th century, historians will say "Nonsense!" as will anyone who was in Homestead, Florida on August 24, 1992.

? **Still Struggling**

In real life, we can't observe infinitely many people and take into account every possible factor in a drug test. We can't toss a die infinitely many times. However, we *can* obtain empirical probability figures that approach "absolute truth" as we continually improve our experimentation methods.

Properties of Outcomes

Let's look at a few formulas that describe properties of outcomes in various types of situations. Don't let the symbology intimidate you. Some of the symbols involve the set theory that you reviewed in Chap. 1.

Law of Large Numbers

Suppose that you toss an "unweighted" die many times. You get numbers turning up, apparently "at random," from the set {1, 2, 3, 4, 5, 6}. What average value will you obtain as you keep tossing the die? If you toss it 100 times, total up the numbers on the faces, and then divide by 100, what will you get? Call this number d (for die). You can reasonably suppose that d will turn out fairly close to the mean, μ. Mathematically, you can say that

$$d \approx \mu$$

where

$$\mu = (1 + 2 + 3 + 4 + 5 + 6)/6$$

$$= 21/6$$

$$= 3.5$$

If you toss a die 100 times, you'll get a value of d that's very close to 3.5. You can expect some variation in the result from one test to another because of "reality imperfection." But now imagine tossing the die 1000 times, or 1,000,000 times, or 1,000,000,000 times! The "reality imperfections" will diminish as you increase the number of tosses. The value of d will converge to 3.5. As you increase the number of tosses without limit, the value of d will come arbitrarily close to 3.5, because the opportunity will diminish for repeated coincidences to skew the result.

The foregoing scenario offers an example of the *law of large numbers*. In a general, informal way, we can state this law as follows:

- As the number of events in an experiment increases, the average value of the outcome approaches the theoretical mean.

This principle constitutes one of the most important laws in probability theory.

Independent Outcomes

We call two outcomes *independent* if and only if the occurrence of one does not affect the probability that the other will occur. We express this criterion symbolically as

$$p(h_1 \wedge h_2) = p(h_1)\, p(h_2)$$

where h_1 and h_2 represent the outcomes, and the inverted V symbol ($\wedge$) stands for the logical connector "and." Figure 3-1 illustrates this scenario in the form of a Venn diagram. The dark-shaded region portrays the intersection of the two sample spaces H_1 and H_2 (light-shaded regions) for an arbitrarily large set of identical experiments in which we observe outcomes h_1 and h_2.

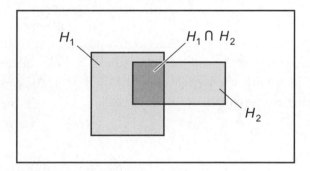

FIGURE 3-1 · Sample spaces H_1 and H_2 for two independent outcomes h_1 and h_2 when we conduct a great many identical experiments.

We can see a good example of independent outcomes when we repeatedly toss a "fair" penny and a "fair" nickel. The face ("heads" or "tails") that turns up on the penny has no effect on the face that turns up on the nickel. It doesn't matter whether we toss the two coins at the same time or at different times. They never interact with each other.

To illustrate how the foregoing formula works in this situation, let $p(p)$ represent the probability that the penny turns up "heads" when we toss a penny and a nickel once. We will see the penny come up "heads" half the time; therefore

$$p(p) = 0.5$$

Let $p(n)$ represent the probability that the nickel turns up "heads" in the same scenario. We'll see the nickel come up "heads" half the time; therefore

$$p(n) = 0.5$$

The probability that both coins will turn up "heads" equals 1 in 4, or 0.25. The foregoing formula states this fact as follows:

$$p(p \wedge n) = p(p)\, p(n)$$

$$= 0.5 \times 0.5$$

$$= 0.25$$

Mutually Exclusive Outcomes

Consider two *mutually exclusive* sample spaces H_1 and H_2. They have no outcomes in common. We can express this situation symbolically as

$$H_1 \cap H_2 = \varnothing$$

In this situation, the probability of either outcome (but not both!) occurring *in a single experiment* equals the sum of their individual probabilities. Here's how we write this fact, with the wedge symbol (∇) representing "either/or," also known as the logical "exclusive-or" operation:

$$p(h_1 \nabla h_2) = p(h_1) + p(h_2)$$

Figure 3-2 illustrates this situation as a Venn diagram. The shaded regions, considered together, portray the union of the two sample spaces H_1 and H_2 for an arbitrarily large set of identical experiments in which we observe outcomes h_1 and h_2.

When two outcomes are mutually exclusive, they can't both occur at the same time. Consider the tossing of a single coin. You'll never see "heads" and "tails"

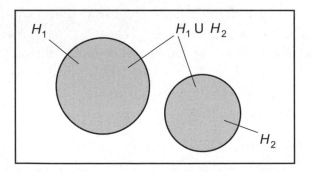

FIGURE 3-2 • Sample spaces H_1 and H_2 for two mutually exclusive outcomes h_1 and h_2 when we conduct a great many identical experiments.

both turn up simultaneously on any given toss. But the sum of the two probabilities (0.5 for "heads" and 0.5 for "tails") equals 1, representing the probability
that one or the other outcome will take place. Of course, we must assume that
the coin never lands on its edge!

As another example, consider the result of a well-run, uncomplicated election for a political office between two candidates. Let's call the candidates
Mrs. Anderson and Mr. Boyd. If Mrs. Anderson wins, we get outcome a, and if
Mr. Boyd wins, we get outcome b. Let's call the respective probabilities of their
winning $p(a)$ and $p(b)$. Imagine that we obtain empirical probability figures by
conducting a poll prior to the election, and we conclude that

$$p_{emp}(a) = 0.29$$

and

$$p_{emp}(b) = 0.71$$

Suppose that, in the event of a tie in the actual election, the applicable laws
prescribe the use of a coin toss to decide the victor! We can now make two
claims with confidence:

- The candidates won't both win.
- One candidate will win.

The probability that either Mrs. Anderson or Mr. Boyd will win equals the sum
of $p(a)$ and $p(b)$, and we know that sum will equal 1 (assuming that neither of
the candidates quits during the election to be replaced by a third, unknown
person, and assuming that we have no write-ins or other election irregularities).
Mathematically, we state this fact as

$$p(a \lor b) = p(a) + p(b)$$
$$= p_{emp}(a) + p_{emp}(b)$$
$$= 0.29 + 0.71$$
$$= 1$$

Complementary Outcomes

Consider two *complementary* sample spaces H_1 and H_2 for an arbitrarily large
set of identical experiments in which we observe outcomes h_1 and h_2 as shown in
Fig. 3-3. (The "drooping minus sign" in front of H_1 at the bottom of the drawing

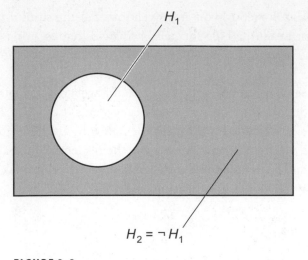

FIGURE 3-3 • Sample spaces H_1 and H_2 for two complementary outcomes h_1 and h_2 when we conduct a great many identical experiments.

translates as the word "not.") We call two outcomes *complementary* if and only if the probability, expressed as a ratio, of one outcome equals 1 minus the probability, expressed as a ratio, of the other outcome. We define this concept in terms of the equations

$$p(h_2) = 1 - p(h_1)$$

and

$$p(h_1) = 1 - p(h_2)$$

Expressed as percentages, we have

$$p_\%(h_2) = 100 - p_\%(h_1)$$

and

$$p_\%(h_1) = 100 - p_\%(h_2)$$

We can exploit the notion of complementary outcomes when we want to calculate the probability that an outcome will *fail* to occur. Consider again the election between Mrs. Anderson and Mr. Boyd. Imagine that you have a "contrarian" attitude and tend to vote *against*, rather than *for*, candidates in elections. You want to know the probability that the candidate you dislike will lose. According to our preelection poll, we found that

$$p_{emp}(a) = 0.29$$

and

$$p_{emp}(b) = 0.71$$

We can rewrite this pair of statements "inside-out" as

$$p_{emp}(\neg a) = p_{emp}(b)$$
$$= 1 - p_{emp}(a)$$
$$= 1 - 0.29$$
$$= 0.71$$

and

$$p_{emp}(\neg b) = p_{emp}(a)$$
$$= 1 - p_{emp}(b)$$
$$= 1 - 0.71$$
$$= 0.29$$

where the "drooping minus sign" ($\neg$) stands for the "not" operation, also called *logical negation*. If you want Mr. Boyd to lose, then you can guess from the pre-election poll that the probability of your wish coming true equals $p_{emp}(\neg b)$, which works out to 0.29.

? Still Struggling

In order for two outcomes to behave in a complementary manner, the sum of their probabilities must equal 1. Therefore, one or the other (but not both) of the two outcomes must take place; they must constitute the only two possible outcomes in a scenario. We can't allow for anomalies such as tossed coins that land on their edges, or elections that end in unbreakable ties.

Nondisjoint Outcomes

We call two outcomes h_1 and h_2 *nondisjoint* if and only if their sample spaces H_1 and H_2 have at least one element in common. Figure 3-4 shows this situation as a Venn diagram. We take away the intersection of the probabilities to ensure that we count the elements common to both sets (represented by the white region where the two sets overlap) only once, not twice.

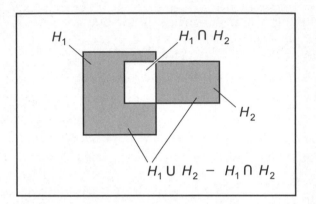

FIGURE 3-4 · Sample spaces H_1 and H_2 for two nondisjoint outcomes h_1 and h_2 when we conduct a great many identical experiments.

With nondisjoint outcomes, the probability of either outcome equals the sum of the probabilities of their occurring separately, minus the probability of their occurring simultaneously. Mathematically, we write this criterion as the formula

$$p(h_1 \vee h_2) = p(h_1) + p(h_2) - p(h_1 \wedge h_2)$$

Here, the wedge symbol $\vee$ represents the logical "or" operation in the inclusive sense, meaning "either or both."

PROBLEM 3-2

Imagine that a certain high school has 1000 students. During his first day on the job, the new swimming and diving coach seeks team prospects. Suppose that the coach has discovered the following facts:

- 200 students can swim well enough to make the swimming team.
- 100 students can dive well enough to make the diving team.
- 30 students can make both teams.

Suppose that the coach wanders through the hallways blindfolded and picks a student "at random." Determine the probabilities, expressed as ratios, that the coach will choose someone who can

- swim fast; call this probability $p(s)$.
- dive well; call this probability $p(d)$.
- swim fast and dive well; call this probability $p(s \wedge d)$.
- either swim fast or dive well, or both; call this probability $p(s \vee d)$.

✔ SOLUTION

Let's assume that the coach has established objective criteria for evaluating prospective candidates for his teams. That said, we must note that the outcomes are not mutually exclusive, nor are they independent. Overlap and interaction exist among the sample spaces. We can find the first three answers immediately, because we've been told the numbers:

$$p(s) = 200/1000$$

$$= 0.200$$

$$p(d) = 100/1000$$

$$= 0.100$$

$$p(s \wedge d) = 30/1000$$

$$= 0.030$$

In order to calculate the last answer—the total number of students who can make either team or both teams—we must find $p(s \vee d)$ using the formula

$$p(s \vee d) = p(s) + p(d) - p(s \wedge d)$$

$$= 0.200 + 0.100 - 0.030$$

$$= 0.270$$

We've determined that 270 (not 300!) students qualify as candidates for either or both teams. The answer would be 300 students if and only if *nobody* had sufficient aquatic skills to make both teams. We mustn't count the exceptional students twice.

Three Mutually Exclusive Outcomes

Suppose that H_1, H_2, and H_3 represent the sample spaces that we get when we perform a large number of experiments with three mutually exclusive outcomes h_1, h_2, and h_3, such that *all three* of the following facts hold true:

$$H_1 \cap H_2 = \varnothing$$
$$H_1 \cap H_3 = \varnothing$$
$$H_2 \cap H_3 = \varnothing$$

In this case, no pair of sample spaces has any elements in common; the three sets are entirely separate from one another. The probability of any one, *but only one*, of the three outcomes occurring equals the sum of their individual probabilities, as follows:

$$p(h_1 \triangledown h_2 \triangledown h_3) = p(h_1) + p(h_2) + p(h_3)$$

We can draw a Venn diagram (Fig. 3-5) to illustrate this situation among sample spaces for an arbitrarily large set of identical experiments in which we observe outcomes h_1, h_2, and h_3.

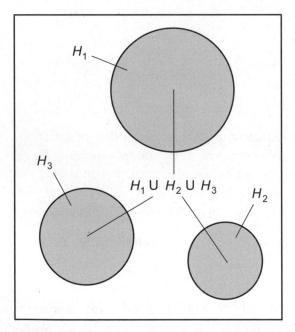

FIGURE 3-5 · Sample spaces H_1, H_2, and H_3 for three mutually exclusive outcomes h_1, h_2, and h_3 when we conduct a great many identical experiments.

Three Nondisjoint Outcomes

Now imagine that H_1, H_2, and H_3 represent the sample spaces that we get when we perform a large number of experiments with three nondisjoint outcomes, so that *one or more* of the following facts holds true:

$$H_1 \cap H_2 \neq \varnothing$$
$$H_1 \cap H_3 \neq \varnothing$$
$$H_2 \cap H_3 \neq \varnothing$$

Here, at least one pair of sample spaces overlaps, sharing one or more elements. The probability that we will observe *at least one* of the outcomes equals the sum of the probabilities of their occurring separately, minus the probabilities of each pair occurring simultaneously, plus the probability of all three occurring simultaneously, as follows:

$$p(h_1 \vee h_2 \vee h_3)$$
$$= p(h_1) + p(h_2) + p(h_3)$$
$$- p(h_1 \wedge h_2) - p(h_1 \wedge h_3) - p(h_2 \wedge h_3)$$
$$+ p(h_1 \wedge h_2 \wedge h_3)$$

We can draw a Venn diagram (Fig. 3-6) to illustrate this situation among sample spaces for an arbitrarily large set of identical experiments in which we observe outcomes h_1, h_2, and h_3.

PROBLEM 3-3

Once again, consider our 1000-student high school with its new aquatics coach. He recruits athletes for the swimming, diving, and water polo teams. Suppose that the following facts hold true among the students:

- 200 people can qualify for the swimming team.
- 100 people can qualify for the diving team.
- 150 people can qualify for the water polo team.
- 30 people can qualify for both the swimming and diving teams.
- 110 people can qualify for both the swimming and water polo teams.
- 20 people can qualify for both the diving and water polo teams.
- 10 people can qualify for all three teams.

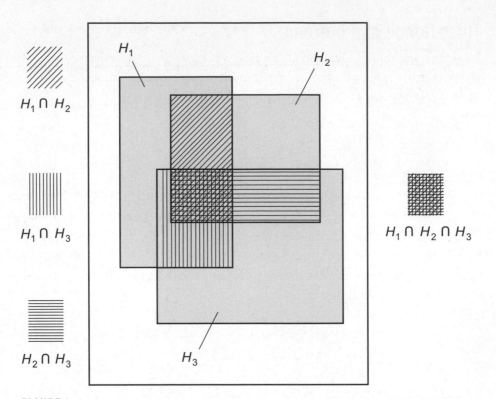

FIGURE 3-6 · Sample spaces H_1, H_2, and H_3 for three nondisjoint outcomes h_1, h_2, and h_3 when we conduct a great many identical experiments.

If the coach chooses students "at random," what's the probability, expressed as a ratio, that he will, on any one choice, select a student good enough for *at least one* of the three water sports?

✓ SOLUTION

Let the following expressions stand for the respective probabilities, all representing the results of random selections by the coach (and all of which we've been told are true):

- Probability that a student can swim fast enough = $p(s)$ = 200/1000 = 0.200
- Probability that a student can dive well enough = $p(d)$ = 100/1000 = 0.100
- Probability that a student can play water polo well enough = $p(w)$ = 150/1000 = 0.150

- Probability that a student can swim and dive well enough $= p(s \wedge d) = $ 30/1000 = 0.030
- Probability that a student can swim and play water polo well enough $= p(s \wedge w) = 110/1000 = 0.110$
- Probability that a student can dive and play water polo well enough $= p(d \wedge w) = 20/1000 = 0.020$
- Probability that a student can swim, dive, and play water polo well enough $= p(s \wedge d \wedge w) = 10/1000 = 0.010$

In order to calculate the total number of students who can effectively perform at least one sport for this coach, we must find $p(s \vee d \vee w)$ as follows:

$p(s \vee d \vee w)$

$= p(s) + p(d) + p(w) - p(s \wedge d) - p(s \wedge w) - p(d \wedge w) + p(s \wedge d \wedge w)$

$= 0.200 + 0.100 + 0.150 - 0.030 - 0.110 - 0.020 + 0.010$

$= 0.300$

In this situation, 300 students constitute good prospects for at least one of the three water sports.

Permutations and Combinations

In probability, we must sometimes choose small sets from large ones, or figure out the number of ways in which certain sets of outcomes can take place. *Permutations* and *combinations* allow us to do these things.

Factorial

We denote the *factorial* of a number by writing an exclamation point after it. For a natural number n greater than or equal to 1, we define $n!$ as the product of all natural numbers less than or equal to n. That is,

$$n! = 1 \times 2 \times 3 \times 4 \times \ldots \times n$$

If $n = 0$, then by default we say that $n! = 1$. We don't define the factorial for any negative number.

As n increases, the value of $n!$ goes up rapidly. When n reaches significant values, the factorial skyrockets. Mathematicians have found a formula for approximating $n!$ when n is large (say, over 50):

$$n! \approx n^n/e^n$$

where e represents a constant called the *natural logarithm base*, an irrational number equal to approximately 2.71828. The "wavy" equals sign emphasizes the fact that the value of $n!$ using this formula represents an approximation, not necessarily an exact figure.

PROBLEM 3-4

Write down the values of the factorial function for $n = 0$ through $n = 15$, in order to illustrate how fast it "blows up."

✔ SOLUTION

Table 3-3 lists the results. You can find these values with a calculator that can display a lot of digits. Most personal computers have calculators good enough for this purpose. If you're lucky, your calculator includes single a key that you can press to get the factorial straightaway!

TABLE 3-3 Values of $n!$ for $n = 0$ through $n = 15$; solution to Problem 3-4.

Value of n	Value of $n!$
0	1
1	1
2	2
3	6
4	24
5	120
6	720
7	5040
8	40,320
9	362,880
10	3,628,800
11	39,916,800
12	479,001,600
13	6,227,020,800
14	87,178,291,200
15	1,307,674,368,000

PROBLEM 3-5

Determine the approximate value of 100! using the formula given above.

SOLUTION

If your calculator has a factorial key, you can solve this problem in a single step. Otherwise you can use a calculator that has an e^x (or natural exponential) function key. In case your calculator can't directly perform either of these two functions, you can find the value of the exponential function using the natural logarithm ("ln") key and the inverse function ("inv") key together. First input x, then hit the "inv" key, and finally hit the "ln" key. The calculator will also need an x^y key (also called $x^\wedge y$) that lets you find the value of any number raised to any power. In addition, the calculator should be capable of displaying numbers in *scientific notation*, also called *power-of-10 notation*. A typical personal-computer calculator will do all of these things if you set it to operate in the "scientific mode." When we apply the foregoing approximate formula for $n = 100$, we get

$$100! \approx (100^{100})/e^{100}$$

$$\approx (10^2)^{100}/(2.688117 \times 10^{43})$$

$$\approx 10^{(2 \times 100)}/(2.688117 \times 10^{43})$$

$$\approx 10^{200}/(2.688117 \times 10^{43})$$

$$\approx 3.72 \times 10^{156}$$

Don't Get Confused!

The numeral representing the value 3.72×10^{156}, if written out in full, would constitute a string of digits too long to fit on most text pages without taking up two or three lines. Your calculator will probably display it as something like

$$3.72e + 156$$

or

$$3.72E156$$

In these displays, the "e" or "E" *does not* represent the natural logarithm base. Instead, it stands for "exponent," in this case meaning "10 raised to the power of."

Permutations

Consider two positive integers q and r. Imagine a set of q items taken r at a time *in a specific order*. We can symbolize the number of different permutations in this situation by writing $_qP_r$ and calculate it with the formula

$$_qP_r = q!/(q-r)!$$

Combinations

Consider two positive integers q and r. Imagine a set of q items taken r at a time *in any order*. We can symbolize the number of different combinations in this situation by writing $_qC_r$ and calculate it with the formula

$$_qC_r = {_qP_r}/r!$$

$$= q!/[r!(q-r)!]$$

PROBLEM 3-6

How many permutations can you obtain with 10 apples, taken five at a time in a specific order?

✔ SOLUTION

Use the above formula for permutations, plugging in $q = 10$ and $r = 5$ to get

$$_{10}P_5 = 10!/(10-5)!$$

$$= 10!/5!$$

$$= 10 \times 9 \times 8 \times 7 \times 6$$

$$= 30{,}240$$

PROBLEM 3-7

How many combinations exist if you have 10 apples, taken five at a time in no particular order?

✔ SOLUTION

Use the above formula for combinations, plugging in $q = 10$ and $r = 5$. You can use the formula that derives combinations based on permutations, because you know from the solution to Problem 3-6 that $_{10}P_5 = 30,240$. Therefore

$$_{10}C_5 = {_{10}P_5}/5!$$

$$= 30,240/120$$

$$= 252$$

The Density Function

When we work with large populations or with continuous random variables, we must define probabilities differently than we do with small populations and discrete random variables. As the number of possible values of a random variable "approaches infinity," we think of the probability that an outcome will fall within a range of values, rather than the probability that an outcome will have any particular value.

A Pattern Emerges

Imagine that some medical researchers want to find out how people's blood pressure levels compare. At first, the researchers select a few dozen individuals and plot the numbers of people having each of 10 specific pressure readings. Suppose they get a graph that looks like Fig. 3-7A. The *systolic pressure* (the higher of the two numbers you get when you take your blood pressure) constitutes the random variable. The points appear to follow a pattern, a result that does not surprise our scientists. They expect most people to have "middling" blood pressure, fewer people to have moderately low or high pressure, and only a small number of people to have extremely low or high blood pressure.

In the next phase of the experiment, the researchers test hundreds of people and specify 20 discrete values of the random variable, getting a plot that looks like Fig. 3-7B. The pattern emerges more clearly. Next, the researchers test thousands of people and plot the results at 40 different blood pressure levels. The resulting plot of frequency (number of people) versus the value of the random variable (blood pressure) reveals a definite pattern now, as shown in Fig. 3-7C. The researchers can rightly assume that repeating the experiment with the same number of subjects (but not the same people) will produce essentially the same pattern every time.

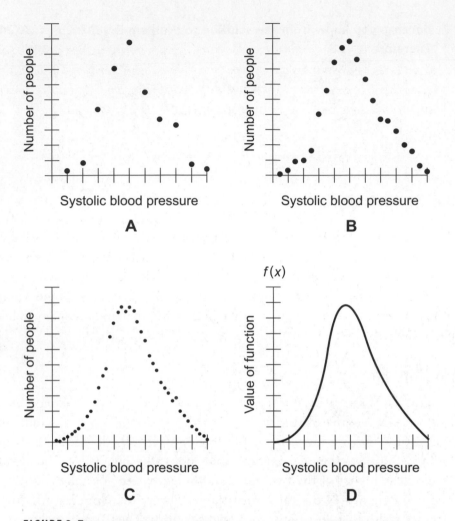

FIGURE 3-7 • Hypothetical plots of blood pressure. At A, a small population and 10 values of pressure; at B, a large population and 20 values of pressure; at C, a gigantic population and 40 values of pressure; at D, the density function.

Expressing the Pattern

Based on the data portrayed in Fig. 3-7C, the researchers can use the technique of curve fitting to derive a general graph for the distribution of blood pressure values, obtaining a continuous curve as illustrated in Fig. 3-7D. This plot shows a *probability density function*, or simply a *density function*. The plot no longer depicts the blood pressure levels of specific individuals. Instead, the curve expresses how blood pressure varies among the general population. We plot the function value $f(x)$, rather than the number of people, along the vertical axis.

TIP *As the number of possible values for the random variable increases without limit, the point-by-point plot blurs into a density function, which we call f (x). The blood pressure of any particular subject vanishes into insignificance. Instead, the researchers concern themselves with the probability that any "randomly" chosen person's blood pressure will fall within a specified range of values.*

Area under the Curve

Figure 3-8 gives us a detailed look at the curve shown in Fig. 3-7D. This function, like all density functions, has a special property: If you calculate or measure the total area under the curve, you always get exactly 1. This rule holds true for the same reason that the relative frequency values of the outcomes for a discrete variable always add up to 1 (or 100%), as we learned in Chap. 2.

Consider two hypothetical systolic blood pressure values: say a and b as shown in Fig. 3-8. We can denote the probability that a "randomly" chosen

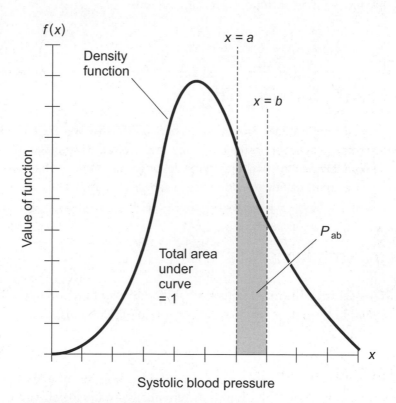

FIGURE 3-8 · The probability that a "randomly" chosen value k of a random variable x will fall between two limit values a and b equals the area under the curve between the vertical lines $x = a$ and $x = b$.

person will have a systolic blood pressure reading k that falls between a and b in any of the following ways:

$$p(a < k < b)$$
$$p(a \leq k < b)$$
$$p(a < k \leq b)$$
$$p(a \leq k \leq b)$$

The first expression includes neither a nor b, the second expression includes a but not b, the third expression includes b but not a, and the fourth expression includes both a and b. All four expressions correspond to the shaded portion of the area under the curve in Fig. 3-8. We can use any of these four interval notations to denote probability. For simplicity and consistency, let's use only expressions of the first type, placing "strictly less-than" signs (<) at both extremes. Then, by convention, we don't include the actual extremes when considering a range of values for a continuous random variable.

? Still Struggling

If we move the vertical lines $x = a$ and $x = b$ back and forth at will, the area of the shaded region gets larger or smaller. In any case, we can never make this area smaller than 0 (when the two lines coincide) or greater than 1 (when the two lines lie so far apart that they encompass the entire area under the curve).

Two Common Distributions

Statisticians describe the nature of a probability density function in the form of a *distribution*. Let's look at two classical types: the *uniform distribution* and the *normal distribution*.

Uniform Distribution

In a uniform distribution, the function remains constant for all values of the random variable. When graphed, a uniform distribution shows up as a flat horizontal line or a rectangle (Fig. 3-9).

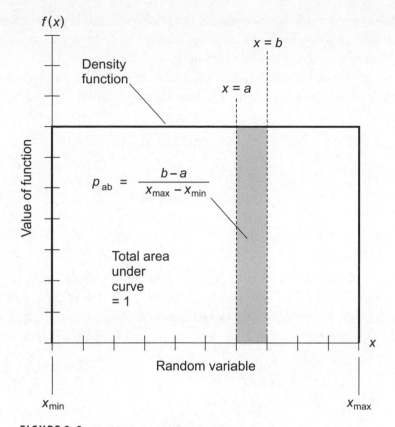

FIGURE 3-9 • A uniform density function has a constant value when the random variable falls between two extremes x_{min} and x_{max}.

Let x represent a continuous random variable. Let x_{min} and x_{max} represent the minimum and maximum values, respectively, that x can attain. In a uniform distribution, x has a density function of the form

$$f(x) = 1/(x_{max} - x_{min})$$

Because the total area under the curve equals 1, the probability p_{ab} that any "randomly" chosen x will fall between a and b is

$$p_{ab} = (b - a)/(x_{max} - x_{min})$$

Imagine that the above-described experiment reveals that equal numbers of people always have each of the given tested blood pressure numbers between two limiting values, say $x_{min} = 100$ and $x_{max} = 140$. Suppose that this result prevails no matter how many people are tested, and no matter how many different values of blood pressure are specified within the range $x_{min} = 100$ to $x_{max} = 140$.

This scenario obviously can't represent the real world, but if you suspend your disbelief and go along with the idea, you can see that this state of affairs produces a uniform probability distribution.

We can calculate the mean (μ), the variance (σ^2), and the standard deviation (σ) for a uniform distribution having a continuous random variable using the formulas

$$\mu = (a + b)/2$$

$$\sigma^2 = (b - a)^2/12$$

$$\sigma = [(b - a)^2/12]^{1/2}$$

Normal Distribution

In a normal distribution, the value of the function has a single central peak and tapers off on either side in a symmetrical fashion. Because of its shape, we call the graph a *bell-shaped curve*. Figure 3-10 shows a generic example. Not all bell-shaped curves portray normal distributions! To qualify as a normal distribution,

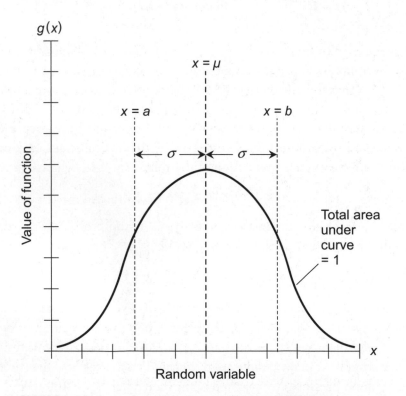

FIGURE 3-10 · The normal distribution, also known as the bell-shaped curve.

a curve must not only have the general shape shown in Fig. 3-10, but must also conform to a rigorous set of rules concerning its standard deviation.

The symbol σ represents the standard deviation of the function, which expresses the extent to which the values of the function concentrate or cluster near the center. It's the same concept that we learned in Chap. 2, but generalized for continuous random variables. A small value of σ produces a "sharp" curve with a narrow peak and steep sides. A large value of σ produces a "broad" curve with less steep sides. As σ approaches 0, the curve narrows toward the mean. If σ grows arbitrarily large, the curve flattens out and settles toward the horizontal axis. In any normal distribution, no matter how "sharp" or "broad," the area under the curve equals 1.

The symbol μ represents the mean. It's the same concept that we learned in Chap. 2, but generalized for continuous random variables. If we want to find the value of μ, we can imagine a moving vertical line that intersects the x axis. When we position the vertical line so that the area under the curve to its left equals 1/2 (or 50%) and the area under the curve to its right equals 1/2 (50%), then our vertical line intersects the x axis at the point $x = \mu$. In a normal distribution, $x = \mu$ at the same point where the function attains its *peak*, or maximum, value.

The Empirical Rule

Refer once again to Fig. 3-10. Imagine two movable vertical lines, one on either side of the vertical line $x = \mu$. Suppose that we position these vertical lines, $x = a$ and $x = b$, so that the one on the left lies at the same distance from $x = \mu$ as the one on the right. The proportion of data points in the part of the distribution $a < x < b$ equals the proportion of the area under the curve between the vertical lines $x = a$ and $x = b$. A well-known theorem in statistics, called the *empirical rule*, states that all normal distributions have the following three characteristics:

- Approximately 68% of the data points lie within the range $\pm\sigma$ of μ.
- Approximately 95% of the data points lie within the range $\pm2\sigma$ of μ.
- Approximately 99.7% of the data points lie within the range $\pm3\sigma$ of μ.

The symbol $\pm$ stands for the words "plus-or-minus."

PROBLEM 3-8

Suppose that you want "mother nature" to provide you with a rainy day to help the plants in your garden grow. It's a gloomy morning. The weather forecasters expect a 50% chance that you'll see up to 1 centimeter (1 cm) of rain come down in the next 24 hours, and a 50% chance that more than 1 cm of rain will fall. They say that you will not, in any event, see more than 2 cm of rain fall, and it's obviously impossible for less than 0 cm of rain to fall. Now imagine that your local radio disc jockeys (DJs) announce the forecast and start talking about a distribution function called $R(x)$ for the rain as predicted by the weather experts. One DJ says that the amount of rain represents a continuous random variable x, and the distribution function $R(x)$ for the precipitation scenario is a normal distribution whose value tails off to 0 at precipitation levels of 0 cm and 2 cm. Draw an approximate distribution graph for this situation.

SOLUTION

Figure 3-11 shows such a plot. The portion of the curve to the left of the vertical line, which represents the mean μ, has an area of 0.5. The mean itself equals 1 cm.

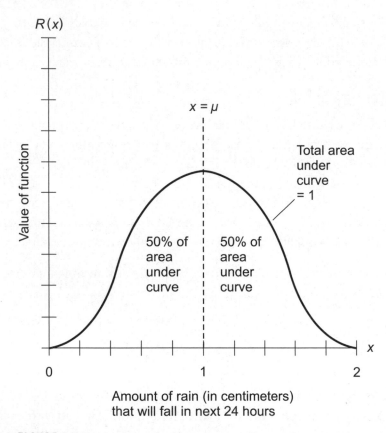

FIGURE 3-11 · Illustration for Problem 3-8.

PROBLEM 3-9

Imagine that in the foregoing scenario, the DJs discuss the extent to which the distribution function "spreads out" around the mean value of 1 cm. One of them mentions a concept called standard deviation, symbolized by a lowercase Greek letter called sigma that looks like a numeral 6 that has tipped over onto its side. Another DJ says that 68% of the total prospective "future wetness" of the town lies in a range of precipitation values defined by sigma on either side of the mean. Draw a crude graph of what the DJs mean by all of this jargon.

SOLUTION

Figure 3-12 illustrates this graph. The shaded region shows the area under the curve between the two vertical lines $x = \mu - \sigma$ and $x = \mu + \sigma$. This region contains 68% of the total area under the curve, centered at the vertical line $x = \mu$ representing the mean.

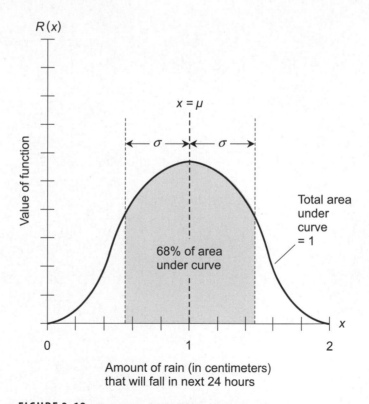

FIGURE 3-12 • Illustration for Problems 3-9 and 3-10.

PROBLEM 3-10

What's the approximate standard deviation in the situation shown by Fig. 3-12?

SOLUTION

The standard deviation appears to be roughly $\sigma = 0.45$, representing the distance of either the vertical line $x = \mu - \sigma$ or the vertical line $x = \mu + \sigma$ from the mean, $x = \mu = 1$ cm.

QUIZ

Refer to the text in this chapter if necessary. A good score is 8 correct. Answers are in the back of the book.

1. In an experiment, if the occurrence of one outcome does not affect the probability that another outcome will occur, we call the two outcomes
   A. disjoint.
   B. independent.
   C. mutually exclusive.
   D. complementary.

2. A single element of a sample space constitutes
   A. an independent variable.
   B. an outcome.
   C. a permutation.
   D. an experiment.

3. Suppose that we know the number C of possible combinations that exist for t objects taken u at a time. We can find the number of permutations straightaway if we
   A. multiply C by t.
   B. multiply C by u.
   C. multiply C by t factorial.
   D. multiply C by u factorial.

4. Based on the appearance of the graph in Fig. 3-13, assuming that both graph scales are linear (meaning that on either axis, every individual division represents the same extent of change in the value of the variable), what's the approximate probability of an outcome falling between $x = a$ and $x = b$?
   A. 1%
   B. 10%
   C. 100%
   D. We can't answer this question without more information.

5. In a situation such as the one shown in Fig. 3-13, we can determine the probability that an outcome will fall between $x = a$ and $x = b$ by
   A. subtracting a from b, and then dividing by 100.
   B. subtracting x_{min} from x_{max}, and then dividing by 100.
   C. subtracting a from b, and then dividing by the quantity $(x_{max} - x_{min})$.
   D. subtracting x_{min} from x_{max}, and then dividing by the quantity $(b - a)$.

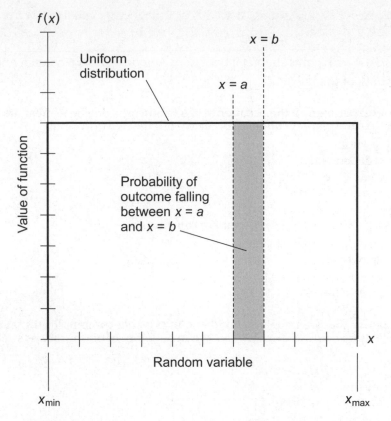

FIGURE 3-13 · Illustration for Quiz Questions 4 and 5.

6. The mathematical probability $p_{math}(x)$, expressed as a ratio or proportion, of a particular outcome x taking place during the course of an experiment always lies between the extremes of
    A. −1 and 1 inclusive.
    B. 0 and 1 inclusive.
    C. −100 and 100 inclusive.
    D. 1 and 100 inclusive.

7. We determine mathematical probability on the basis of
    A. theoretical calculations.
    B. the results of a single experiment.
    C. the results of many experiments.
    D. the variance and the standard deviation.

8. How many possible permutations exist for six objects taken five at a time?
   A. 6
   B. 30
   C. 720
   D. 900

9. How many possible combinations exist for six objects taken five at a time?
   A. 6
   B. 30
   C. 720
   D. 900

10. We define an *event* as
   A. one test in an experiment.
   B. one outcome in an experiment.
   C. a set of two or more outcomes in an experiment.
   D. the set of all tests in an experiment.

chapter 4

Descriptive Measures

When we analyze data, we can break it down into intervals, graph it in various ways, or describe its characteristics as numbers derived from formulas. Statisticians call these techniques *descriptive measures*.

CHAPTER OBJECTIVES

In this chapter, you will

- Define and determine percentiles, quartiles, and deciles
- Generate interval data from tables and graphs
- Construct nomographs and pie graphs
- Learn to use fixed-width and variable-width histograms
- Define and determine ranges, coefficients of variation, and Z scores
- Calculate interquartile ranges

Percentiles

When you attended elementary school, did you take standardized tests every year? In my time (the 1960s) and place (Minnesota), all grammar-school students took the "Iowa Tests of Basic Skills." At the end of one of those ordeals, the teacher said that I had scored in a certain *percentile*. Then she told me what that meant!

Percentiles in a Normal Distribution

Statisticians can use the percentile method to divide a large data set into 100 intervals, each interval containing 1% of the elements in the set. A total of 99 (not 100) percentiles exist, because we define each percentile according to the boundary point between two adjacent intervals.

Imagine an experiment in which we obtain the *systolic blood pressure* readings for a large number of people. (The systolic reading constitutes the larger of two numbers that the measuring instrument, called a *sphygmomanometer*, displays.) If you have a blood pressure of 110/70, read "110 over 70," then your systolic pressure equals 110. Suppose that we obtain the results of this experiment in graphical form for a huge population, and we see the normal distribution shown in Fig. 4-1.

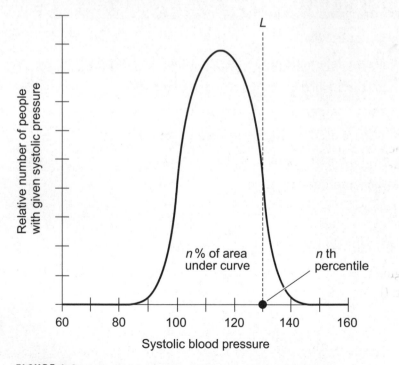

FIGURE 4-1 • Determination of percentiles in a normal distribution.

Let's choose a systolic pressure value on the horizontal axis, and extend a line L straight up from that point. We can determine the percentile corresponding to this value by finding the number n such that at least $n\%$ of the area under the curve falls to the left of line L. Then, we round n off to the nearest whole number between, and including, 1 and 99 to get the percentile ranking p. For example, suppose that the region to the left of the line L represents 93.3% of the area under the curve. Therefore, $n = 93.3$. Rounding to the nearest whole number between 1 and 99 yields the percentile value $p = 93$. Now we know that the systolic blood pressure corresponding to the point where line L intersects the horizontal axis lies in the 93rd percentile.

We can determine any particular percentile point (boundary), say the pth, by locating the vertical line such that the percentage n of the area beneath the curve to the left of line L exactly equals p and then noting the point where line L crosses the horizontal axis. Imagine that we can move the line L in Fig. 4-1 to the left and right at will. When the number n, representing the percentage of the area beneath the curve to the left of L, equals precisely 93, then the line L crosses the horizontal axis at the 93rd percentile boundary point. Although it's tempting to think that a "zeroth percentile" $(n = 0)$ and a "100th percentile" $(n = 100)$ exist, neither of these notions corresponds to a boundary where two adjacent intervals meet.

? Still Struggling

Note the difference between saying that a certain pressure "is in" the pth percentile, versus saying that a certain pressure "is at" the pth percentile. In the former case, we describe a data interval. In the latter case, we describe a boundary point between two intervals.

Percentiles in Tabular Data

Imagine that 1000 students take a 40-question test. A total of 41 possible scores exist: 0 through 40 inclusive. Suppose that at least one of the students gets a score corresponding to every element of the set $\{0, 1, 2, 3, ..., 38, 39, 40\}$. Therefore, every score is accounted for. Some smart (and lucky) people write perfect papers, and a few unfortunates fail to get any of the answers right. Table 4-1 shows the test results, with the scores in ascending order from 0 to 40 in the

TABLE 4-1 Results of a hypothetical 40-question test taken by 1000 students.

Test Score	Absolute Frequency	Cumulative Absolute Frequency
0	5	5
1	5	10
2	10	20
3	14	34
4	16	50
5	16	66
6	18	84
7	16	100
8	12	112
9	17	129
10	16	145
11	16	161
12	17	178
13	22	200
14	13	213
15	19	232
16	18	250
17	25	275
18	25	300
19	27	327
20	33	360
21	40	400
22	35	435
23	30	465
24	35	500
25	31	531
26	34	565
27	35	600
28	34	634
29	33	667
30	33	700
31	50	750

(Continued)

TABLE 4-1	Results of a hypothetical 40-question test taken by 1000 students. (*Continued*)	
Test Score	**Absolute Frequency**	**Cumulative Absolute Frequency**
32	50	800
33	45	845
34	27	872
35	28	900
36	30	930
37	28	958
38	20	978
39	12	990
40	10	1000

first (left-most) column. For each possible score, the number of students getting that score (the absolute frequency) appears in the second column. The third (right-most) column shows the cumulative absolute frequency, expressed from lowest to highest scores.

Where do we put the 99 percentile points (boundaries) in this data set? Obviously, we can't insert 99 boundary points into a set that has only 41 possible values! However, we can try a different approach to the conundrum of figuring out the percentile rankings. We know that 1000 individual students have taken the test. We can partition the set of students into 100 different groups with 99 different boundaries, and then call the 99 boundaries the percentile points as follows:

- The "worst" 10 papers, and the first percentile point at the top of that group
- The "second worst" 10 papers, and the second percentile point at the top of that group
- The "third worst" 10 papers, and the third percentile point at the top of that group

↓

- The "pth worst" 10 papers, and the pth percentile point at the top of that group

↓

- The "$(100 - n)$th best" 10 papers, and $(100 - n)$th percentile point at the bottom of that group

↓

- The "third best" 10 papers, and the 97th percentile point at the bottom of that group
- The "second best" 10 papers, and the 98th percentile point at the bottom of that group
- The "best" 10 papers, and the 99th percentile point at the bottom of that group

This scheme seems like a great idea at first, but we have a problem. When we check Table 4-1, we can see that 50 people scored 31 on the test. We have five groups of 10 people, all with the same score. These scores are all "equally good." If we intend to say that *any one* of these papers falls into the *p*th percentile, then we must say that they *all* fall into the *p*th percentile. We shouldn't arbitrarily take 10 papers with scores of 31 and put them into the *p*th percentile, then take 10 more papers with scores of 31 and put them in the $(p + 1)$st percentile, then take 10 more papers with scores of 31 and put them in the $(p + 2)$nd percentile, then take 10 more papers with scores of 31 and put them into the $(p + 3)$rd percentile, and finally take 10 more papers with scores of 31 and put them into the $(p + 4)$th percentile. If we did that, then we'd treat some students more favorably than others, and we should expect that the "losers" would protest!

Percentile Points

We can avoid the foregoing complications by defining a scheme for calculating the positions of the percentile points in a set of *ranked data elements*. A set of ranked data elements constitutes a set arranged in a table from "worst to best" such as Table 4-1. Once we have defined the percentile positioning scheme, we accept it as a convention (standard operating procedure).

Imagine that we want to find the position of the *p*th percentile boundary point in a set of *n* ranked data elements. First, we multiply *p* by *n*, and then we divide the product by 100. This calculation gives us a number *i* called the *index*, such that

$$i = pn/100$$

If *i* doesn't equal a whole number, then we define the *p*th percentile point as $i + 1$. If *i* equals a whole number, then we define the *p*th percentile point as $i + 0.5$.

Percentile Ranks

If we want to find the percentile *p* for a given element or position *s* in a ranked data set, we use a different definition. We divide the number of elements less than s (call this number t) by the total number of elements n, and multiply this quantity by 100, getting a "tentative percentile" p^* such that

$$p^* = 100t/n$$

Then we round p^* to the nearest whole number between, and including, 1 and 99 to get the percentile rank p for that element or position in the set.

When we define percentiles in this way, we obtain intervals whose centers lie at the percentile boundaries as defined earlier. The first and 99th percentiles often turn out slightly "oversized" according to this scheme, especially if we have a large population. That effect takes place because the first and 99th percentile ranks encompass the elements at the very extremes of a set or distribution. We call such elements *outliers*.

Don't Get Confused!

Once in a while you'll hear people use the term "percentile" in an inverted, or upside-down, sense. They'll talk about the "first percentile" when they really mean the 99th, the "second percentile" when they really mean the 98th, and so on. Beware of this conceptual glitch! If you get done with a test and you think that you scored high, and then you're told that you're in the "fourth percentile," don't panic. Ask the teacher or test administrator, "What does that mean, exactly? The top 4%? The top 3%? The top 3.5%? Or what?"

PROBLEM 4-1

Where does the 56th percentile point lie in the data set shown by Table 4-1?

SOLUTION

We have a total of 1000 students (data elements), so $n = 1000$. We want to find the 56th percentile point, so $p = 56$. First, let's calculate the index:

$$i = (56 \times 1000)/100$$

$$= 56{,}000/100$$

$$= 560$$

This result equals a whole number, so we must add 0.5 to it, getting $i + 0.5 = 560.5$. Now we know that the 56th percentile constitutes the boundary between the "560th worst" and 561st worst" test papers. To find out what score this represents, we must check the cumulative absolute frequency values in Table 4-1. The cumulative frequency corresponding to a score of 25 equals 531 (that's less than 560.5); the cumulative frequency corresponding to a score of 26 equals 565 (that's more than 560.5). The 56th percentile point therefore lies between the scores of 25 and 26.

PROBLEM 4-2

If you take the above-described test and get a score of 33 correct answers, what's your percentile rank?

SOLUTION

Checking the table, you can see that 800 students have scores less than yours. (Not less than or equal to, but *strictly* less than!) According to the second definition given above, $t = 800$. The tentative percentile $p*$ is therefore

$$p* = 100 \times 800/1000$$

$$= 100 \times 0.8$$

$$= 80$$

This equals a whole number, so rounding it to the nearest whole number between 1 and 99 yields $p = 80$. You scored in the 80th percentile.

PROBLEM 4-3

If you take the test and score 0, what's your percentile rank?

SOLUTION

In this case, no one has a score lower than yours. Therefore, according to the second definition given above, $t = 0$. The tentative percentile $p*$ is

$$p* = 100 \times 0/1000$$

$$= 100 \times 0$$

$$= 0$$

Remember, you must round to the nearest whole number between 1 and 99 to get the actual percentile value. In this case, that number is $p = 1$. Therefore, you rank in the first percentile.

Quartiles and Deciles

We can use other methods besides the percentile scheme to divide up data sets. Statisticians sometimes specify points or boundaries that divide data into quarters or into tenths.

Quartiles in a Normal Distribution

We can define a *quartile* or *quartile point* as any of three numbers that break a data set into four intervals, each interval containing approximately 1/4 or 25% of the elements in the set. Three quartiles, not four, exist because the quartiles represent boundaries where adjacent intervals meet. Therefore, we assign quartiles the values 1, 2, or 3 only, calling them the *first quartile*, the *second quartile*, and the *third quartile*.

Let's examine Fig. 4-1 again. We can find the qth quartile point by locating the vertical line L such that the percentage n of the area beneath the curve to the left of L equals exactly $25q$ and then noting the point where L crosses the horizontal axis. Imagine that we can move line L freely back and forth, to the left or to the right, in Fig. 4-1. Let n represent the percentage of the area under the curve that lies to the left of L. The quartiles appear as follows:

- When $n = 25\%$, line L crosses the horizontal axis at the first quartile point.
- When $n = 50\%$, line L crosses the horizontal axis at the second quartile point.
- When $n = 75\%$, line L crosses the horizontal axis at the third quartile point.

Quartiles in Tabular Data

Let's return to the 40-question test described above and in Table 4-1. Where do we put the three quartile points in this data set? There are 41 different possible scores and 1000 actual data elements. We can break these 1000 results up into four different groups with three different boundary points according to the following criteria:

- The highest possible boundary point representing the "worst" 250 or fewer papers, and the first quartile point at the top of that set
- The highest possible boundary point representing the "worst" 500 or fewer papers, and the second quartile point at the top of that set
- The highest possible boundary point representing the "worst" 750 or fewer papers, and the third quartile point at the top of that set

Figure 4-2A is a *nomograph* (a diagram that compares numerical scales side by side) that illustrates the positions of the quartile points for the test results shown by Table 4-1.

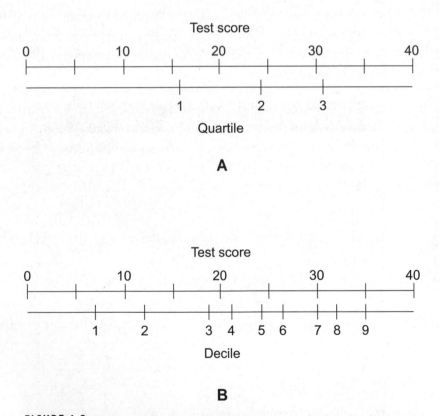

A

B

FIGURE 4-2 · At A, positions of quartiles in the test results described in the text. At B, positions of the deciles.

? Still Struggling

The data in Table 4-1 are unusual. They represent a coincidence because we can clearly define all three quartile points. Obvious boundaries exist between the "worst" 250 papers and the "second worst," between the "second and third worst," and between the "third worst" and the "best." These boundaries occur at the transitions between scores of 16 and 17, 24 and 25, and 31 and 32 for the first, second, and third quartiles, respectively. If we give these same 1000 students another 40-question test, or if we administer this same 40-question test to a different group of 1000 students, we should not expect to have such good luck with the quartile points.

PROBLEM 4-4

Table 4-2 shows a portion of results for the same 40-question test as the one described above, but with slightly different results from those shown in Table 4-1, so that we can't "cleanly" define the first quartile point. Where's the first quartile point here?

TABLE 4-2 Table for Problem 4-4.		
Test Score	**Absolute Frequency**	**Cumulative Absolute Frequency**
↑	↑	↑
↑	↑	↑
↑	↑	↑
13	22	200
14	13	213
15	19	232
16	16	248
17	30	278
18	22	300
19	27	327
↓	↓	↓
↓	↓	↓
↓	↓	↓

✔️ **SOLUTION** _____

Interpret the definition literally. The first quartile represents the *highest possible* boundary point at the top of the set of the "worst" 250 *or fewer* papers. In Table 4-2, that corresponds to the transition between scores of 16 and 17.

Deciles in a Normal Distribution

We can define a *decile* or *decile point* as any of nine numbers that break a data set into 10 intervals, each interval containing about 1/10 or 10% of the elements in the set. The deciles represent the points where the 10 sets meet. We assign deciles whole-number values between, and including, 1 and 9, calling them the first decile, the second decile, the third decile, and so on up to the ninth decile.

Refer again to Fig. 4-1. We can find the dth decile point by locating the vertical line L such that the percentage n of the area beneath the curve to the left of L equals $10d$ and then noting the point where L crosses the horizontal axis. Imagine again that we can slide L to the left and right at will. Let n represent the percentage of the area under the curve that lies to the left of L. Then the decile points appear as follows:

- When $n = 10\%$, L crosses the horizontal axis at the first decile point.
- When $n = 20\%$, L crosses the horizontal axis at the second decile point.
- When $n = 30\%$, L crosses the horizontal axis at the third decile point.

↓

- When $n = 90\%$, line L crosses the horizontal axis at the ninth decile point.

Deciles in Tabular Data

One more time, let's scrutinize the 40-question test whose results appear in Table 4-1. Where do we put the decile points? We break the 1000 test papers into 10 different groups with nine different boundary points according to the following criteria:

- The highest possible boundary point representing the "worst" 100 or fewer papers, and the first decile point at the top of that set
- The highest possible boundary point representing the "worst" 200 or fewer papers, and the second decile point at the top of that set
- The highest possible boundary point representing the "worst" 300 or fewer papers, and the third decile point at the top of that set

↓

- The highest possible boundary point representing the "worst" 900 or fewer papers, and the ninth decile point at the top of that set

The nomograph of Fig. 4-2B on page 116 illustrates the positions of the decile points for the test results shown by Table 4-1. As with the quartiles, clear boundaries exist between the "worst" 100 papers and the "second worst," between the "second and third worst," between the "third and fourth worst," and so on up.

TIP *If we give these same 1000 students a different 40-question test, or if we give this 40-question test to a different group of 1000 students, we should expect that the locations of the decile points will not turn out as "clean" as they do in Table 4-1 and Fig. 4-2B. (By now, some of you will suspect that I contrived Table 4-1 to make things come out "clean." You're right.)*

PROBLEM 4-5

Table 4-3 shows a portion of a 40-question test with slightly different results from those portrayed in Table 4-1. Here, you can't "cleanly" define the sixth decile point in terms of whole-number scores. Where should you put that point?

TABLE 4-3 Table for Problem 4-5.		
Test Score	**Absolute Frequency**	**Cumulative Absolute Frequency**
↑	↑	↑
↑	↑	↑
↑	↑	↑
24	35	500
25	31	531
26	34	565
27	37	602
28	32	634
29	33	667
30	33	700
↓	↓	↓
↓	↓	↓
↓	↓	↓

✔️ **SOLUTION**

Once again, interpret the definition literally. The sixth decile constitutes the *highest possible* boundary point at the top of the set of the "worst" 600 *or fewer* papers. In Table 4-3, that corresponds to the transition between scores of 26 and 27.

Intervals by Element Quantity

Percentiles, quartiles, and deciles can get confusing when you hear a statement such as, "You're in the 99th percentile of this graduating class. That's the highest possible rank." Doubtlessly, more than one student in this elite class has asked, upon being told such a thing, "Don't you mean that I'm in the 100th percentile? After all, the term 'percentile' implies that 100 groups should exist, not 99." That's a good question and comment. In fact, if a teacher tells you that you are "in" a such-and-such percentile, she makes a technically imprecise statement. Your score can fall "in" an interval, but not "in" a boundary point between intervals!

Boundaries versus Interval Centers

You can think in terms of the intervals between percentile boundaries, quartile boundaries, or decile boundaries, rather than the intervals centered at the boundaries. In fact, from a purely mathematical standpoint, this approach makes more sense, as follows:

- The 99 percentile points in a ranked data set divide the set into 100 intervals, each of which has an equal number (or as nearly an equal number as possible) of elements.
- The nine decile points divide a ranked data set into 10 intervals as nearly equal-sized as possible.
- The three quartile points divide a ranked set into four intervals as nearly equal-sized as possible.

25% Intervals

Let's return to Table 4-1 and imagine that we want to express the scores in terms of the lowest 25%, the second lowest 25%, the second highest 25%, and the highest 25%. Table 4-4 shows the test results with the 25% intervals portrayed.

TABLE 4-4 Results of a hypothetical 40-question test taken by 1000 students, with the 25% intervals indicated.

Range of Scores	Absolute Frequency	Cumulative Absolute Frequency	25% Intervals
0–16	250	250	Lowest 25%
17–24	250	500	Second lowest 25%
25–31	250	750	Second highest 25%
32–40	250	1000	Highest 25%

We can also call these intervals the first quarter, the second quarter, the third quarter, and the fourth quarter.

10% Intervals

Now suppose that we want to express the scores in Table 4-1 according to the lowest 10%, the second lowest 10%, the third lowest 10%, and so forth up to the highest 10%. Table 4-5 shows the test results with the 10% intervals portrayed. We can also call these spans the first 10th, the second 10th, the third 10th, the fourth 10th, and so on up to the tenth 10th (i.e., the highest 10th).

TABLE 4-5 Results of a hypothetical 40-question test taken by 1000 students, with the 10% intervals indicated.

Range of Scores	Absolute Frequency	Cumulative Absolute Frequency	10% Intervals
0–7	100	10	Lowest 10%
8–13	100	200	Second lowest 10%
14–18	100	300	Third lowest 10%
19–21	100	400	Fourth lowest 10%
22–24	100	500	Fifth lowest 10%
25–27	100	600	Fifth highest 10%
28–30	100	700	Fourth highest 10%
31–32	100	800	Third highest 10%
33–35	100	900	Second highest 10%
36–40	100	1000	Highest 10%

TIP   *As you already know, this particular set of scores constitutes a special case, because the interval cutoff points appear "clean." If I hadn't contrived this set to make the discussion so easy, we'd have had to do more work to find the quartile or decile points, and then define the 25% or 10% intervals as the sets of scores between those boundaries.*

PROBLEM 4-6

Table 4-6 shows a portion of results for a 40-question test given to 1000 students, but with slightly different results from those portrayed in Table 4-1. What range of scores represents the second highest 10% (i.e., the ninth 10th) this instance?

TABLE 4-6   Table for Problem 4-6.

Test Score	Absolute Frequency	Cumulative Absolute Frequency
↑	↑	↑
↑	↑	↑
↑	↑	↑
30	35	702
31	51	753
32	50	803
33	40	843
34	27	870
35	31	901
36	30	931
↓	↓	↓
↓	↓	↓
↓	↓	↓

✔ SOLUTION

The second highest 10% corresponds to the range of scores bounded at the bottom by the eighth decile and at the top by the ninth decile. The eighth decile constitutes the *highest possible* boundary point at the top of the set of the "worst" 800 *or fewer* papers. In Table 4-6, that point corresponds to the transition between scores of 31 and 32. The ninth decile is the *highest possible* boundary point at the top of the set of the "worst" 900 *or fewer* papers. In Table 4-6, that point corresponds to the transition between scores of 34 and 35. The second highest 10% (or the ninth 10th) therefore falls in the range of scores from 32 to 34, inclusive.

PROBLEM 4-7

Table 4-7 shows a portion of results for a 40-question test given to 1000 students, but with different results from those portrayed in Table 4-1. What range of scores represents the lowest 25% in this instance?

TABLE 4-7	Table for Problem 4-7.	
Test Score	**Absolute Frequency**	**Cumulative Absolute Frequency**
↑	↑	↑
↑	↑	↑
↑	↑	↑
14	12	212
15	18	230
16	19	249
17	27	276
18	26	302
↓	↓	↓
↓	↓	↓
↓	↓	↓

SOLUTION

The lowest 25% corresponds to the range of scores bounded at the bottom by the lowest possible score and at the top by the first quartile. The lowest possible score is 0. The first quartile is the *highest possible* boundary point at the top of the set of the "worst" 250 *or fewer* papers. In Table 4-7, that point corresponds to the transition between scores of 16 and 17. The lowest 25% therefore constitutes the range of scores from 0 to 16, inclusive.

Fixed Intervals

In this chapter so far, we've split data sets into subsets containing equal (or as nearly equal as possible) numbers of elements, and then observed the ranges of values that fall into each subset. Alternatively, we can define fixed ranges of independent-variable values, and then observe the number of elements in each range.

The Test Revisited

Let's reexamine the test whose results appear in Table 4-1 and think about the ranges of scores. Tables 4-8, 4-9, and 4-10 denote three distinct methods of evaluating the ranges.

In Table 4-8, we lay out the results of the test according to the number of papers having scores in the ranges 0–10, 11–20, 21–30, and 31–40. We see that the largest number of students scored in the range 21–30, followed by the ranges 31–40, 11–20, and 0–10.

In Table 4-9, we portray the test results according to the number of papers having scores in 10 ranges. In this case, the most "popular" score range is 29–32. The next most "popular" range is 21–24. The least "popular" range is 0–4.

Tables 4-8 and 4-9 both divide the test scores into equal-sized ranges (except the lowest range, which includes one extra score, the score of 0). In Table 4-10, we tabulate the scores according to letter grades A, B, C, D, and F that the teacher (a stern pedagogue indeed) sees fit to assign.

TABLE 4-8   Results of a hypothetical 40-question test taken by 1000 students, divided into four equal ranges of scores.

Range of Scores	Absolute Frequency	Percentage of Scores
0–10	145	14.5
11–20	215	21.5
21–30	340	34.0
31–40	300	30.0

TABLE 4-9 Results of a hypothetical 40-question test taken by 1000 students, divided into 10 equal ranges of scores.

Range of Scores	Absolute Frequency	Percentage of Scores
0–4	50	5.0
5–8	62	6.2
9–12	66	6.6
13–16	72	7.2
17–20	110	11.0
21–24	140	14.0
25–28	134	13.4
29–32	166	16.6
33–36	130	13.0
37–40	70	7.0

TABLE 4-10 Results of a hypothetical 40-question test taken by 1000 students, divided into ranges of scores according to subjective grades.

Letter Grade	Range of Scores	Absolute Frequency	Percentage of Scores
F	0–18	300	30.0
D	19–24	200	20.0
C	25–31	250	25.0
B	32–37	208	20.8
A	38–40	42	4.2

Pie Graph

We can portray the data in any of the three Tables 4-8, 4-9, or 4-10 in graphical form using broken-up circles to create a *pie graph*, also called a *pie chart*. When we want to generate a graph of this sort, we divide a circle into wedge-shaped sections, exactly as we would slice up a round pizza pie to prepare snacks for several people. As the size of the data subset increases, the *angular width* or *apex angle* of the pie section increases in direct proportion.

In Fig. 4-3, graph A portrays the data results from Table 4-8, graph B portrays the results from Table 4-9, and graph C portrays the results from Table 4-10.

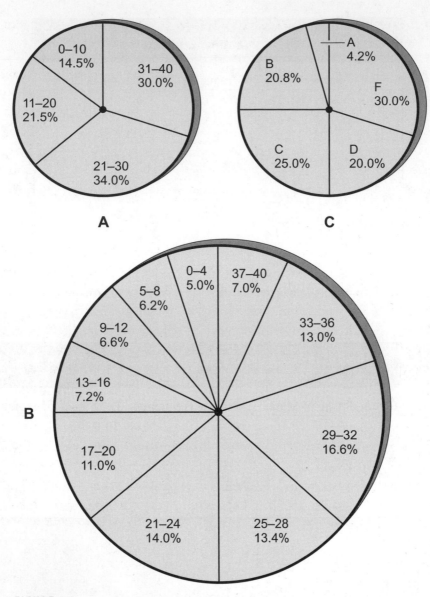

FIGURE 4-3 · At A, pie graph of data in Table 4-8. At B, pie graph of data in Table 4-9. At C, pie graph of data in Table 4-10.

The angle at the apex of each wedge (at the center of the pie), in degrees, depends directly on the percentage of data elements in the subset, as follows:

- If a wedge portrays 10% of the students, its apex angle measures 10% of 360°, or 36°

- If a wedge portrays 25% of the students, its apex angle measures 25% of 360°, or 90°

- In general, if a wedge portrays x% of the elements in the population, the apex angle of the wedge, in degrees, equals $3.6x$

As an alternative to the apex-angle scheme, we can express sizes of the wedges of each pie in terms of the proportion or percentage of the area inside the circle. The wedges all have the same radius—equal to the radius of the circle—so their areas vary in direct proportion to the percentages of the data elements in the subsets they portray. For example, in Fig. 4-3A, the range of scores 31–40 corresponds to a slice containing 30% or 3/10 of the pie, while in Fig. 4-3C, the set of students who got grades of D corresponds to a slice containing 20% or 1/5 of the pie.

Variable-Width Histogram

We took a brief look at the concept of the histogram in Chap. 1. The example shown in that chapter represented a *fixed-width histogram*. There exists a more flexible type of histogram, called the *variable-width histogram*, which can portray situations such as the results of the hypothetical 40-question test given to 1000 students in diverse ways. Variable-width histograms usually appear as vertical bar graphs.

Figure 4-4 shows histograms that express the same data as that in the tables and pie graphs that we just finished examining. Graph A portrays the data results from Table 4-8, graph B portrays the results from Table 4-9, and graph C portrays the results from Table 4-10 as a variable-width histogram. The widths of the vertical bars vary in direct proportion to the range of numerical test scores. In these instances, the only histogram where the bars actually vary in width is the one shown at C, for the letter grades, because that's the only case where the range of scores varies from bar to bar. The heights of the bars go in direct proportion to the percentage of students who received scores in the indicated range.

? Still Struggling

We've labeled the percentages at the tops of the bars in Fig. 4-4A because we have room enough to show the numbers without cluttering up the presentation. In Fig. 4-4B and C, we haven't written the percentages at the top of each bar. Showing the numbers in graph B would make it look congested. In graph C, showing the percentage for the grade of A at the top of that bar would be nigh impossible, so we haven't listed any of the values. We can always clarify histograms by including the tabular data along with the graphs.

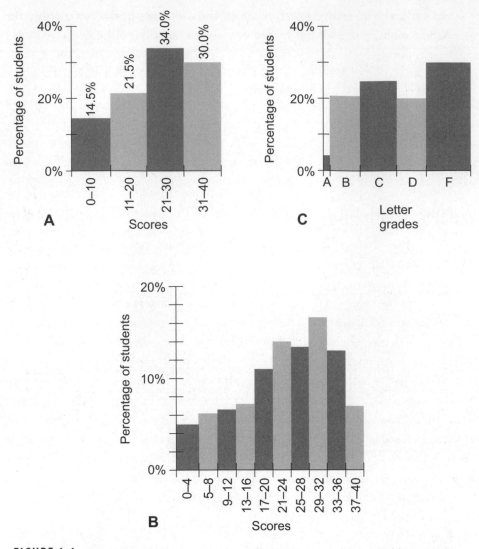

FIGURE 4-4 • At A, histogram of data in Table 4-8. At B, histogram of data in Table 4-9. At C, histogram of data in Table 4-10.

PROBLEM 4-8

Imagine a large corporation that operates on a 5-day workweek (Monday through Friday). Suppose that we average the number of workers who call in sick each day of the week over a period of many weeks, and we average the number of sick-person-days per week over the same period. (A sick-person-day represents the equivalent of one person staying home sick for 1 day. If the same person calls in sick for 3 days in a given week, we get 3 sick-person-days for that week, even though we have only one sick person.) For each day of the workweek, we divide the average number of people who call in sick on that day by the average number of sick-person-days per week, and then tabulate the result as a percentage for that day. Figure 4-5 shows the results as a pie graph. Name two significant things that this graph tells us about Fridays. Name one thing that this graph might at first seem to, but actually does not, tell us about Fridays.

✔ SOLUTION

The pie graph of Fig. 4-5 indicates that more people (on the average) call in sick on Fridays than on any other day of the workweek. The graph also tells us that, of the total number of sick-person-days on a weekly basis, an average of 33.3% of them occur on Fridays. The pie graph might at first seem to, but actually doesn't, indicate that an average of 33.3% of the workers in the corporation call in sick on Fridays.

PROBLEM 4-9

Suppose that, in the above-described corporation and over the survey period portrayed by Fig. 4-5, we observe 1000 sick-person-days per week

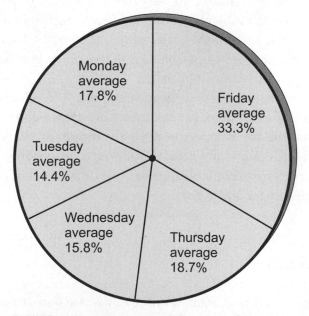

FIGURE 4-5 • Illustration for Problems 4-8 through 4-10.

on average. What's the average number of sick-person-days on Mondays? What's the average number of people who call in sick on Mondays?

✔ SOLUTION

For any single day, a sick-person-day represents the equivalent of one person calling in sick. But this definition does not necessarily apply to a period longer than 1 day. In this single-day example, we can multiply 1000 by 17.8% to obtain 178. This result gives us both answers. On the "average Monday" we see 178 sick-person-days. In other words, an average of 178 individuals call in sick on Mondays, considered over the course of many weeks.

PROBLEM 4-10

Given the same scenario as that described in the previous two problems, what's the average number of sick-person-days on Mondays and Tuesdays combined (i.e., on a Monday or a Tuesday)? What's the average number of individuals who call in sick on Mondays and Tuesdays combined?

✔ SOLUTION

An average of 178 sick-person-days fall on a Monday, as we determined in the solution to Problem 4-9. To find the average number of sick-person-days that fall on a Tuesday, we multiply 1000 by 14.4%, getting 144. The average number of sick-person-days that occur on Mondays and Tuesdays combined equals 178 + 144, or 322.

We can't determine the average number of individuals who call in sick on Mondays and Tuesdays combined, because we don't know how many Monday-Tuesday sick-person-day pairs represent one single individual staying out sick on both days, as opposed to how many Monday-Tuesday sick-person-per-day pairs represent two different people, each of whom stays out sick for only 1 day.

Other Specifications

We can take advantage of additional descriptive measures to evaluate the characteristics of data. Let's look at three of the most common parameters.

Range

In a data set, or in any *contiguous* (unbroken) interval in that set, we define the *range* as the difference between the smallest value and the largest value in the set or interval.

In the graph of hypothetical blood-pressure test results (Fig. 4-1), the lowest systolic pressure in the data set is 60 and the highest is 160. The range therefore equals 160 − 60, or 100. If any of the people whom we tested happened to exhibit systolic blood pressures lower than 60 or higher than 160, we simply discarded them for the purpose of generating Fig. 4-1.

In the 40-question test we've examined numerous times in this chapter, the lowest score is 0 and the highest score is 40. The range therefore equals 40 − 0, or 40. If we want, we can restrict our attention to only some scores in a single contiguous interval, for example the second lowest 25% of the scores. We can determine that range by examining Table 4-4; it happens to equal 24 − 17, or 7.

? Still Struggling

In this context, the meaning of the term *range* differs slightly from the meaning of the same term at the top of the left-hand column in Table 4-4. In the table, one of the ranges constitutes the *span* of scores from 17 through 24; here, we consider the *difference* between the span extremes, or the single number 7. Both of these meanings differ substantially from the definition of the term *range* relevant to mathematical functions in calculus or analysis.

Coefficient of Variation

Do you remember the definitions of the mean (μ) and the standard deviation (σ) from Chap. 2? Let's review them briefly. We can derive an important descriptive measure from them.

In a normal distribution, such as the one that shows the results of our hypothetical blood-pressure data-gathering experiment, the mean equals the value (in this case the blood pressure) such that we observe the same area under the curve on either side of a vertical line corresponding to that value. In tabulated data for discrete elements, the mean equals the arithmetic average of all the results.

In a normal distribution, the standard deviation expresses the extent to which the data appears "spread-out" around the mean. As the data becomes more "spread-out," the curve grows "flatter" and the standard deviation increases. In tabulated data for discrete elements, if we have results $\{x_1, x_2, x_3, ..., x_n\}$ whose mean equals μ, then we can calculate the standard deviation using the formula

$$\sigma = \{(1/n)[(x_1 - \mu)^2 + (x_2 - \mu)^2 + \cdots + (x_n - \mu)^2]\}^{1/2}$$

In any case, the standard deviation expresses the extent of dispersion or "spread-outedness," while the mean expresses the location of the average or center point.

Now suppose that we want to know how "spread-out" the data appears, *relative to the mean*. We can get an expression for this parameter if we divide the standard deviation by the mean. We call the resulting ratio the *coefficient of variation*, symbolized as CV. Mathematically, we have

$$CV = \sigma/\mu$$

We express the standard deviation and the mean in the same units, such as systolic blood pressure or test score. When we divide one parameter by the other, the units in the denominator cancel out the units in the numerator, so the CV doesn't need any unit whatsoever. It's simply a number. We call a quantity that lacks any associated unit a *dimensionless quantity*.

Because CV is dimensionless, we can use it to compare the "spread-outedness" of data sets that describe vastly different phenomena or effects, such as blood pressures and test scores. A large CV means that data appears "spread out a lot" around the mean. A small CV means that data appears concentrated closely around the mean. In the extreme, if CV = 0, all of the data values equal the mean! Figure 4-6 shows two distributions in graphical form, one with a low CV and the other with a much higher CV.

TIP   *A potential difficulty exists in conjunction with the above formula. Have you guessed it? If you wonder what happens in a distribution where the data can attain either positive or negative values—for example, temperatures in degrees Celsius—your concern is justified. If $\mu = 0$ (the freezing point of water on the Celsius temperature scale), we have a problem. We can avoid this conundrum by changing the units in which we specify the data, so that 0 doesn't occur within the set of possible values. When expressing temperatures, we can use the Kelvin scale where the readings never drop below 0. In a situation where all the*

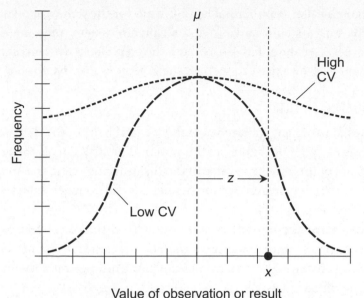

FIGURE 4-6 • Two distributions shown in graphical form, one with a low coefficient of variation (CV) and one with a higher CV. Derivation of the Z score (z) for result x is discussed in the text.

elements in a data set equal 0, such as would happen if a whole class of students turns in blank papers on a test, we can't define the CV because the mean actually equals 0.

Z Score

Sometimes, you'll hear people say that a such-and-such observation or result lies "2.2 standard deviations below the mean" or "1.6 standard deviations above the mean." The Z *score*, symbolized z, gives us a quantitative measure of the position of a particular element with respect to the mean. The Z score of an element equals the number of standard deviations by which the element differs from the mean, either positively or negatively.

For a specific element x in a data set, the value of z depends on both the mean (μ) and the standard deviation (σ). We can find it with the formula

$$z = (x - \mu)/\sigma$$

If x lies below the mean, then z turns out as a negative number ($z < 0$). If x lies above the mean, then z is a positive number ($z > 0$). If x equals the mean, then $z = 0$.

In the graphical distributions of Fig. 4-6, $z > 0$ for the point x shown. This situation holds true for both curves. We can't precisely determine the Z score for either curve by looking at Fig. 4-6, because the graph lacks the necessary numeric information. However, we can see that the Z score is positive for both curves.

Interquartile Range

Sometimes, statisticians want to know the "central half" of the data in a set. The *interquartile range* (IQR) expresses this notion. The IQR equals the value of the third quartile point minus the value of the first quartile point. If a quartile point occurs between two integers, we can define it as the average of the two integers (the smaller one plus 0.5).

Consider again the hypothetical 40-question test taken by 1000 students. The quartile points appear in Fig. 4-2A. The first quartile point lies between the scores of 16 and 17; the third quartile point lies between the scores of 31 and 32. Therefore

$$IQR = 31 - 16$$
$$= 15$$

PROBLEM 4-11

Suppose that we administer an entirely new 40-question test to 1000 students, different from any of the other tests described earlier in this chapter. The scores turn out much more closely concentrated than those from the test depicted in Fig. 4-2A. How would the IQR of this test compare with the IQR of the previous test?

SOLUTION

We would observe a smaller IQR, because the first and third quartile points would lie closer together.

PROBLEM 4-12

The empirical rule, which we learned in Chap. 3, states that all normal distributions have the following three characteristics:

- Approximately 68% of the data points lie within the range $\pm\sigma$ of μ.
- Approximately 95% of the data points lie within the range $\pm 2\sigma$ of μ.
- Approximately 99.7% of the data points lie within the range $\pm 3\sigma$ of μ.

Restate this principle in terms of Z scores.

✔ SOLUTION

As defined above, the Z score of an element equals the number of standard deviations that the element departs from the mean, either positively or negatively. Therefore, according to the empirical rule, all normal distributions have the following three characteristics:

- Approximately 68% of the data points have Z scores between –1 and +1.
- Approximately 95% of the data points have Z scores between –2 and +2.
- Approximately 99.7% of the data points have Z scores between –3 and +3.

QUIZ

Refer to the text in this chapter if necessary. A good score is 8 correct. Answers are in the back of the book.

1. Suppose that some students take a 20-question quiz. The least successful students get only two answers correct, while the best students can manage 18 answers correct. What's the range, expressed as a single number?

    A. 9
    B. 10
    C. 16
    D. 18

2. Suppose that 80 students take a 20-question quiz. The third quartile point is

    A. the point located at a score of 15, which represents 3/4 of 20.
    B. the point located at a score of 5, which represents 1/4 of 20.
    C. the lowest point representing the "best" 60 or more papers.
    D. the highest point representing the "worst" 60 or fewer papers.

3. Imagine that a great many students take a 100-question quiz, and the mean turns out to equal exactly 66 answers correct. Suppose that we analyze the data further and get a figure of $\sigma = 13.2$. What's the coefficient of variation?

    A. We need more information to calculate it.
    B. 52.8
    C. 5.0
    D. 0.20

4. Suppose that several students take a 10-question quiz. The worst score is 3 correct, and the best score is 10 correct. What's the Z score for this test?

    A. 7, which equals $10 - 3$
    B. 6.5, which equals $(3 + 10)/2$
    C. $30^{1/2}$, which equals the square root of (3×10)
    D. None of the above

5. Imagine that you see a pie graph portraying the results of a survey designed to determine the number and proportion of families in Happytown earning incomes in various ranges. The pie graph does not show actual numbers, but one of the ranges has a "slice" that constitutes exactly half of the "pie." From this fact, you can assume that the "half-pie" slice corresponds to

    A. the families in Happytown whose earnings fall in the interquartile range.
    B. the upper-income half of the families surveyed in Happytown.
    C. the lower-income half of the families surveyed in Happytown.
    D. None of the above

6. Table 4-11 shows the results of a hypothetical 10-question test given to 140 students. The second decile point lies

TABLE 4-11 Results of a hypothetical 10-question test given to 140 students. Table for Quiz Questions 6 through 10.

Test Score	Absolute Frequency	Cumulative Absolute Frequency
0	2	2
1	1	3
2	2	5
3	5	10
4	8	18
5	22	40
6	42	82
7	30	112
8	11	123
9	10	133
10	7	140

A. at the transition between scores of 2 and 3.
B. at the transition between scores of 4 and 5.
C. at the transition between scores of 7 and 8.
D. in a position that isn't clear from this table.

7. What's the range for the scores achieved by students in the scenario of Table 4-11, expressed as a single number?

A. 5
B. 8
C. 10
D. We can't determine it without more information.

8. Where's the 10th percentile point for the test results shown in Table 4-11?

A. At the transition between scores of 3 and 4
B. At the transition between scores of 5 and 6
C. Precisely at the score of 9
D. It doesn't exist because the notion is undefined.

9. Suppose, in the scenario shown by Table 4-11, Mary M. takes the test and gets eight answers correct. In what interval does Mary M's score fall with respect to the class?

   A. The highest 10%
   B. The second highest 10%
   C. The third highest 10%
   D. The fourth highest 10%

10. What's the interquartile range for the test results portrayed in Table 4-11?

   A. 2
   B. 3
   C. 4
   D. We can't determine it because the quartile points don't fall on exact score values.

Test: Part I

Do not refer to the text when taking this test. You may draw diagrams or use a calculator if necessary. A good score is at least 45 correct. Answers are in the back of the book. It's best to have a friend check your score the first time, so you won't memorize the answers if you want to take the test again.

1. A graph that shows proportions as vertical bars having variable width and height constitutes an example of a
   A. lateral graph.
   B. pie graph.
   C. histogram.
   D. point-set spreadsheet.
   E. nomograph.

2. Fill in the blank to make the following sentence true: "Deciles divide a data set into intervals, each interval containing about _____ of the elements in the set."
   A. 11%
   B. 10%
   C. 9%
   D. 5%
   E. 1%

3. Suppose that two 20-element sets share 10 elements in common. From this information, we can deduce that the union of the two sets must contain
   A. no elements.
   B. 10 elements.
   C. 20 elements.
   D. 30 elements.
   E. 40 elements.

4. Figure Test I-1 shows the results of research in five imaginary towns, compiled on the first day of May, 2010. This illustration is an example of a

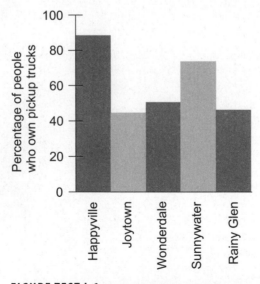

FIGURE TEST I-1 • Illustration for Part I Test Questions 4 through 6.

A. proportional line graph.
B. point-to-point graph.
C. correlation chart.
D. cumulative frequency graph.
E. vertical bar graph.

5. **What, if anything, is technically wrong with Fig. Test I-1?**
   A. The bars should get taller as we move toward the left.
   B. The bars should get taller as we move toward the right.
   C. Actual numbers of people should appear at the tops of the bars.
   D. The bars should vary in width as well as in height.
   E. Nothing is technically wrong with this graph.

6. **If we erase all the bars in Fig. Test I-1 and then change the values on the vertical axis to represent the actual number of people in each town who own pickup trucks (rather than percentages), what must we do before we can draw in the new vertical bars for the data?**
   A. Determine the total population all five towns combined.
   B. Reduce all the numbers by a factor of 100.
   C. Ensure that all the numbers add up to 100.
   D. Increase all the numbers by a factor of 100.
   E. None of the above

7. **We can illustrate the unions and intersections of multiple sets by drawing a**
   A. pie graph.
   B. bar graph.
   C. Venn diagram.
   D. histogram.
   E. regional graph.

8. **When two variables bear no correlation whatsoever and we plot their values as points on a coordinate grid, the points appear**
   A. along a well-defined line or curve.
   B. clustered near the top of the grid.
   C. clustered near the bottom of the grid.
   D. clustered near the center of the grid.
   E. scattered all over the grid.

9. **When we observe a number of events in an experiment, we call the results**
   A. outcomes.
   B. independent variables.
   C. dependent variables.
   D. random variables.
   E. Z scores.

10. **A subset of a population constitutes**

    A. an experiment.
    B. a variable.
    C. a sample.
    D. an element.
    E. an outcome.

11. **A quadratic equation always has at least**

    A. one real-number solution.
    B. two real-number solutions.
    C. three real-number solutions.
    D. four real-number solutions.
    E. None of the above

12. **What type of graph best portrays a normal distribution for a data set containing millions of elements or a continuous range of values?**

    A. A scatter plot
    B. A continuous-curve graph
    C. A histogram
    D. A bar graph
    E. An interquartile plot

13. **Which of the following constitutes an example of a discrete random variable?**

    A. The temperature outside your house
    B. The speed of a car on a highway
    C. The amount of sunlight falling on a solar panel
    D. The number of dots that turn up when you toss a gambling die
    E. The thrust of a rocket engine during liftoff

14. **Imagine that a large number of students take a test, producing results as shown in Fig. Test I-2. This illustration constitutes an example of a**

    A. point-to-point graph.
    B. continuous-curve graph.
    C. histogram.
    D. bar graph.
    E. normal distribution.

15. **In a graph of the type shown in Fig. Test I-2, we should ensure that the sum of the five dependent-variable values never exceeds**

    A. 1/5 of 100%, or 20%.
    B. 1/2 of 100%, or 50%.
    C. 4/5 of 100%, or 80%.
    D. 100%.
    E. the value of the ninth decile point.

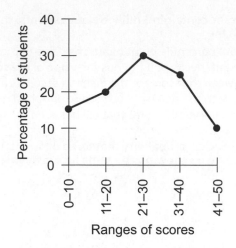

FIGURE TEST I-2 • Illustration for Part I Test Questions 14 through 16.

16. **In a graph of the type shown in Fig. Test I-2, what's the smallest possible dependent-variable value to which a point could correspond?**

    A. 0%
    B. 10%
    C. 15%
    D. 30%
    E. We need more information to answer this question.

17. **Suppose that we have a bucket full of paper scraps with multi-digit numerals written on them. A mathematician tells us that some of the numerals are complicated, but they all represent rational numbers. Based on that information, we know with *absolute certainty*, before we pick a tag from the bucket, that the *numerical value* of the numeral on the tag will be expressible as**

    A. a positive or negative whole number.
    B. a positive whole number, a negative whole number, or zero.
    C. an integer divided by a positive whole number.
    D. an integer divided by zero.
    E. the product of two integers.

18. **In a ranked data set containing millions of elements, the second quartile point appears in essentially the same place as the**

    A. first decile point.
    B. ninth decile point.
    C. first percentile point.
    D. 99th percentile point.
    E. 50th percentile point.

19. In a ranked data set containing millions of elements, the interquartile range essentially equals the

    A. value of the 75th percentile point minus the value of the 25th percentile point.

    B. value of the eighth decile point minus the value of the second decile point.

    C. difference between the scores at the first and second quartile points.

    D. difference between the scores at the second and third quartile points.

    E. average of the numbers of scores that fall between the quartile points.

20. What's the mathematical probability that three different "unweighted" 12-faced "superdice" will come up showing the same face if we toss them all at once?

    A. 1 in 12

    B. 1 in 144

    C. 1 in 1728

    D. 1 in 20,376

    E. 1 in 531,441

21. How many 5% intervals exist in a set of 260 ranked data elements?

    A. 52

    B. 26

    C. 20

    D. 13

    E. We need more information to answer this question.

22. When we manipulate an equation, we can perform any of the following actions *except one*, and get another equation representing the same situation as the original equation does. Which action produces a potentially inaccurate or useless result?

    A. Multiply each side by zero.

    B. Add the same constant to both sides.

    C. Add one of the variables in question to both sides.

    D. Multiply each side by one of the variables in question.

    E. Add a quantity that can attain a value of zero to both sides.

23. Which, if any, of the following statements A, B, C, or D holds true for the graph shown in Fig. Test I-3?

    A. Curve R has a lower CV than curve U.

    B. Curve U has a lower CV than curve R.

    C. Line S lies at the first quartile.

    D. Line T portrays the mode.

    E. None of the above statements holds true for the situation portrayed in Fig. Test I-3.

24. Suppose that line S lies precisely at the mean for curve R in Fig. Test I-3. In that case, and based on the information given in the graph, we know that line T lies

    A. at the median for curve R.

    B. at the mode for curve R.

    C. at the variance point for curve R.

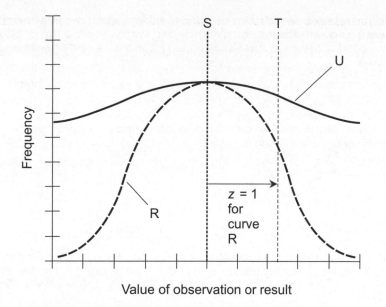

Value of observation or result

FIGURE TEST I-3 · Illustration for Part I Test Questions 23 through 25.

D. one standard deviation above the mean for curve R.

E. one decile above the mean for curve R.

25. **Suppose that curve R in Fig. Test I-3 portrays a normal distribution. In that case, the area of the region bounded on the left by line S, on top by curve R, on the right by line T, and on the bottom by the horizontal axis represents approximately**

A. 17% of the data.

B. 25% of the data.

C. 34% of the data.

D. 40% of the data.

E. 45% of the data.

26. **Let *q* represent a set of items or objects taken *r* at a time in no particular order, where *q* and *r* represent positive integers. We symbolize the possible number of combinations in this situation as $_qC_r$, and we can calculate it using the formula**

$$_qC_r = q!/[r!(q-r)!]$$

Given this information, what's the possible number of combinations of seven objects taken four at a time?

A. 5040

B. 840

C. 210

D. 42

E. 35

27. Let q represent a set of items or objects taken r at a time in a specific order, where q and r represent positive integers. We symbolize the possible number of permutations in this situation as $_qP_r$, and we can calculate it using the formula

$$_qP_r = q!/(q-r)!$$

Given this information, what's the possible number of permutations of seven objects taken four at a time?

A. 5040
B. 840
C. 210
D. 42
E. 35

28. Imagine two nonempty sets called A and B that have some elements in common. Suppose that x represents one of those elements. We can have absolute confidence that one and only one of the following statements is false. Which one?

A. $x \in A \cup B$
B. $x \in A \cap B$
C. $x \notin A$
D. $x \in B$
E. $x \in A$

29. The likelihood of a particular outcome for an event, based on experience or observation, is called

A. statistical probability.
B. mathematical probability.
C. numerical probability.
D. empirical probability.
E. discrete probability.

30. Fill in the blank to make the following sentence true: "In a nondecreasing function, the value of the dependent variable never grows any _____ as the value of the independent variable increases."

A. more positive or less negative
B. more negative or less positive
C. less constant
D. more random
E. less chaotic

31. Fill in the blank to make the following sentence true: "In a distribution that appears as a smooth curve on a graph, the _____ is the value x such that, if we construct a vertical line L_x intersecting the independent-variable axis at x, the area under the curve to the left of L_x equals the area under the curve to the right of L_x."

A. mean
B. mode
C. standard deviation
D. Z score
E. variance

32. **When a sample comprises an entire population, we call it**
    A. an outcome.
    B. a range.
    C. an independent variable.
    D. a function.
    E. a census.

33. **If you complete a standardized test and the instructor tells you that your score lies between the 95th and 96th percentiles for your class, you can conclude that your score**
    A. is among the lowest in the class.
    B. is among the highest in the class.
    C. is near the middle of the class.
    D. has a high coefficient of variation.
    E. has a high mean.

34. **Suppose that the regions X and Y in Fig. Test I-4 represent two different sets of outcomes in an experiment. The pair of light-shaded regions (excluding the dark-shaded zone) portrays the set of outcomes belonging to**
    A. both X and Y.
    B. neither X nor Y.
    C. either X or Y, but not both.
    D. X or Y, or both.
    E. X but not Y.

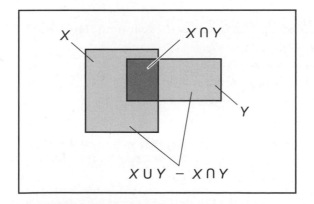

FIGURE TEST I-4 • Illustration for Part I Test Questions 34 and 35.

35. In Fig. Test I-4, the white (completely unshaded) region outside both small rect-
     angles represents the set of outcomes belonging to
     A. both X and Y.
     B. neither X nor Y.
     C. either X or Y, but not both.
     D. X or Y, or both.
     E. X but not Y.

36. What do we call the set of all possible outcomes in the course of an experiment?
     A. An independent variable
     B. A universe
     C. A sample space
     D. A dependent variable
     E. A population

37. Fill in the blank to make the following statement true: "A relation can behave as
     a function *if and only if* every individual independent-variable value corresponds
     to _____ dependent-variable value."
     A. exactly one
     B. at least one
     C. at least two
     D. infinitely many
     E. at most one

38. Suppose that if an element x belongs to a nonempty set Q, then x cannot, *under
     any circumstances*, belong to another nonempty set P. From this fact, we can
     conclude that
     A. P constitutes a subset of Q.
     B. P constitutes a member of Q.
     C. Q constitutes a subset of P.
     D. Q constitutes a member of P.
     E. P and Q are disjoint.

39. For a large positive integer, one of the following statements is true. Which one?
     A. $n! = n/(n-1)$
     B. $n! = n \times (n-1) \times (n-2) \times (n-3) \ldots \times 3 \times 2 \times 1$
     C. $n! = n + (n-1) + (n-2) + (n-3) \ldots + 3 + 2 + 1$
     D. $n! = n/(n+1)$
     E. $n! = n^{(n-1)}$

40. What's the mathematical probability that an "unweighted" die, tossed twice, will
     show the face with four dots on the first toss and the face with two dots on the
     second toss?
     A. 1 in 6
     B. 1 in 18
     C. 1 in 36

D. 1 in 512

E. 1 in 1296

41. What's the mathematical probability that an "unweighted" die, tossed twice, will show the face with four dots on one toss and two dots on the other toss (either four dots and then two dots, or two dots and then four dots)?

A. 1 in 6

B. 1 in 18

C. 1 in 36

D. 1 in 512

E. 1 in 1296

42. Consider two nonempty sets called C and D. Suppose that set $C \subseteq D$ and, at the same time, $D \subseteq C$. What can we say about the sets?

A. They're disjoint.

B. Their union contains no elements.

C. Their intersection is identical to their union.

D. Set C is the complement of set D.

E. Set D is the complement of set C.

43. Consider a function in which time constitutes the independent variable and temperature constitutes the dependent variable. We see a graph of this function with the temperature plotted at 15-minute intervals for the duration of one complete 24-hour day, except that the data between 1:00 p.m. and 3:00 p.m. is missing. We fill in this data on the graph by connecting the 1:00 and 3:00 points with a straight line, and hope that the results closely portray the actual situation for that day. What "educated guesswork" tactic have we used?

A. Histography

B. Functional extrapolation

C. Median plotting

D. Linear interpolation

E. Mode estimation

44. Table Test I-1 shows the results of an experiment in which we toss a single gambling die 100 times. The "X factor" in this table constitutes the

TABLE TEST I-1   Table for Part I Test Question 44.

Face of Die	Absolute Frequency	X Factor
1	17	17
2	16	33
3	17	50
4	18	68
5	14	82
6	18	100

A. cumulative absolute frequency.
B. variant frequency.
C. mathematical frequency.
D. summation frequency.
E. dependent frequency.

45. Table Test I-2 shows the results of an experiment in which we have tossed a single gambling die, manufactured by a different vendor than the die used in the preceding experiment (Table Test I-1). What can we reasonably suspect about the dice in these experiments?

TABLE TEST I-2	Table for Part I Test Question 45.	
Face of Die	Absolute Frequency	X Factor
1	10	10
2	27	37
3	15	52
4	8	60
5	10	70
6	30	100

A. Both dice are "unweighted."
B. Both dice are "weighted."
C. The die we tossed in the previous experiment (Table Test I-1) is "unweighted," while the die we tossed in this experiment (Table Test I-2) is "weighted."
D. The die we tossed in the previous experiment (Table Test I-1) is "weighted," while the die we tossed in this experiment (Table Test I-2) is "unweighted."
E. Nothing in particular.

46. Fill in the blank in the following sentence to make it true: "In a normal distribution, a _____ defines any one of 99 values or points that divides a data set into intervals, each interval containing an equal proportion of the elements in the data set."
A. Z score
B. median
C. decile
D. boundary
E. percentile

47. **We call the "middle value" for the outcomes in an experiment, such the number of outcomes at or above it equals the number of outcomes at or below it, the**
   A. median.
   B. mode.
   C. Z score.
   D. mean.
   E. central tendency.

48. **Figure Test I-5 shows a general illustration of**
   A. a normal distribution.
   B. an invariant distribution.
   C. a random distribution.
   D. a uniform distribution.
   E. a variant distribution.

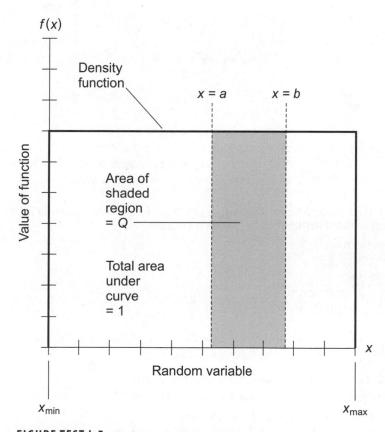

FIGURE TEST I-5 · Illustration for Part I Test Questions 48 through 50.

49. In Fig. Test I-5, the area Q of the shaded region between the two dashed vertical lines portrays
    A. the standard deviation for the span of x-values between a and b.
    B. the mean for the span of x-values between a and b.
    C. the Z score for the span of x-values between a and b.
    D. the probability that any "randomly" chosen x will fall between a and b.
    E. the variance range for the span of x-values between a and b.

50. Which of the following statements about Fig. Test I-5 is true?
    A. $Q = (b + a)/(x_{max} \, x_{min})$
    B. $Q = (b - a)/(x_{max} - x_{min})$
    C. $Q = (b - a)/(x_{max} + x_{min})$
    D. $Q = ba/(x_{max} \, x_{min})$
    E. We need more information to calculate Q.

51. Which of the following mathematical probability figures $p(k)$ for a given outcome k in a sample space K represents an impossible scenario?
    A. $p(k) = 0.0$
    B. $p(k) = 2/3$
    C. $p(k) = 1.0$
    D. $1/(2^{1/2})$
    E. $p(k) = 2^{1/2}$

52. In any contiguous interval in a data set, the range equals the
    A. average of all the values in the interval.
    B. largest value minus the smallest value in the interval.
    C. average of the largest and smallest values in the interval.
    D. value such that equally many other values lie above and below it in the interval.
    E. value or values that we observe most often in the interval.

53. Suppose that in a Venn diagram, a circle and its interior portray a set called S. A single point within that circle can represent
    A. an element of S.
    B. a subset of S.
    C. the union of S.
    D. the intersection S.
    E. the complement of S.

54. Imagine that a certain high school has 500 students. The football and baseball coach has determined the following three facts:

    • 80 students can qualify for the football team.
    • 60 students can qualify for the baseball team.
    • 20 students can qualify for both teams.

Suppose that the coach selects one student "at random." What's the probability that the coach will choose a student who can qualify for the football team?

A. 4%

B. 8%

C. 12%

D. 16%

E. 20%

55. In the scenario of Question 54, what's the probability that the coach will "randomly" choose a student who can qualify for the baseball team?

A. 4%

B. 8%

C. 12%

D. 16%

E. 20%

56. In the scenario of Question 54, what's the probability that the coach will "randomly" choose a student who can qualify for both the football team and the baseball team?

A. 4%

B. 8%

C. 12%

D. 16%

E. 20%

57. The variance and the standard deviation both provide us with expressions of the

A. number of outcomes that we get when we do an experiment.

B. skewing of data toward either extreme in a bar graph or point-to-point graph.

C. extent to which the data appears "spread-out" on either side of the mean.

D. maximum extent to which we can allow error in our calculations.

E. maximum error that actually occurs in experimental observations.

58. In a normal distribution, at which decile point does the second quartile point lie?

A. The second.

B. The third.

C. The fourth.

D. The fifth.

E. We need more information to answer this question.

59. How many quartile points exist a set of 60 ranked data elements?

A. 240

B. 15

C. Four

D. Three

E. None, because quartile points can't exist in sets of ranked data elements

60. **The mode for a discrete variable tells us the**
    A. average of all the values.
    B. value such that equally many other values lie above and below it.
    C. value or values that we observe most often.
    D. extent to which the values appear "spread-out."
    E. extent to which the values appear "clustered together."

Part II

Statistics in Action

chapter 5

Sampling and Estimation

In the real world, we gather data by *sampling*, and determine its characteristics by *estimation*. These processes require care and precision to minimize errors and avoid misinterpretation.

CHAPTER OBJECTIVES

In this chapter, you will

- Learn to gather and organize primary or secondary source data
- Interpret information revealed by data
- Evaluate sampling processes
- Visually interpolate data displayed on analog devices
- Minimize experimental error
- Learn the Central Limit Theorem
- Define and determine confidence intervals

Source Data and Sampling Frames

When we conduct a statistical experiment, we should execute four steps in the following sequence.

1. Formulate the question. What do we want to know, and what (or who) do we want to learn about?

2. Gather the data from the required places, from the right people, and over a sufficient period of time.

3. Organize and analyze the data so that it becomes *information*.

4. Interpret the information gathered and organized from the experiment, thereby obtaining *knowledge*.

Primary versus Secondary

If we have no data for analysis, we must collect data on our own. When a statistician gathers data herself, she calls it *primary source data*. If data are already available and the statistician has only to organize and analyze it, she calls it *secondary source data*.

In the case of primary source data, we must follow the proper collection schemes, and then we must use the proper methods to organize, evaluate, and interpret it. That is, we must carry out each of the above steps 2, 3, and 4 correctly. With secondary source data, the collection process has already been done for us, but we still have to organize, evaluate, and interpret it, so we have to execute steps 3 and 4.

Either way, we face plenty of potential pitfalls. Many things can go wrong with an experiment, but there's rarely more than one way to get it right.

Sampling Frames

The most common data-collection schemes involve obtaining samples that represent populations with little or no *bias*, meaning that our only agenda should comprise revealing the truth (as opposed to supporting or "propping up" someone's belief or opinion). With small populations, we usually find that task easy, because we can sample the entire population. However, with populations too large and diverse to sample in their entirety, an unbiased sampling scheme can prove difficult to conceive and carry out.

In Chap. 2, we learned that the term *population* refers to a particular set of items, objects, phenomena, or people undergoing analysis, such as the set of all the insects in the world. We also defined the term *sample* in Chap. 2. A sample of a population constitutes a subset of that population, for example, the set of all the mosquitoes in the world that carry the malaria protozoan.

In some situations, we'll want to define a set intermediate between a sample and a population. This is often the case with gigantic populations. A *sampling frame* is a set of items within a population from which we choose a sample. We want to "whittle down" the size of the sample, while still obtaining a fair representation of the population. In the mosquito experiment, the sampling frame might be the set of all mosquitoes caught by a team of researchers, one for each 10,000 square kilometers of land surface area in the world, on the first day of each month for one complete calendar year. We could then test all the recovered insects for the presence of the malaria protozoan.

In the simplest case, the sampling frame coincides with the population as shown in Fig. 5-1A. However, in the mosquito experiment described above, the sampling frame is small in comparison with the population (Fig. 5-1B). Occasionally, a population is so large, diverse, and complicated that we'll want to use two sampling frames, one inside the other, as shown in Fig. 5-1C. If the number of mosquitoes caught in the above process is so large that it would take too much time to individually test them all, we could select, say, 1% of the mosquitoes "at random" from the ones caught, and then test the individual insects one by one.

Choosing Frames

The choice of sampling frames represents an important consideration whenever we undertake to conduct a statistical experiment. Each sampling frame must constitute a fair (unbiased) representation of the population if we expect the results of the experiment to portray reality!

Imagine that we want to evaluate some particular characteristic of real numbers. Therefore, our population is the set of all real numbers. We can choose various sampling frames from this vast population, among them the following:

- The set of all positive integers
- The set of all integers
- The set of all rational numbers
- The set of all irrational numbers
- The set of all transcendental numbers

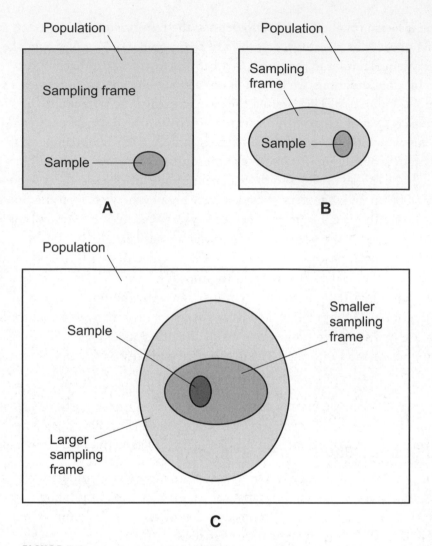

FIGURE 5-1 • At A, the sampling frame coincides with the population. At B, the sampling frame constitutes a proper subset of the population. At C, two sampling frames exist, one inside the other.

Suppose that we choose the set of all rational numbers as the sampling frame. Within this set, we can specify subframes such as the following:

- The set of all integers
- The set of all even integers
- The set of all odd integers
- The set of all integers divisible by 5

- The set of all rational numbers that aren't integers
- The set of all rational numbers whose numerals turn out infinitely long when we try to write them in decimal form

Finally, within this subframe, we choose a sample. How about the set of integers divisible by 100? Or the set of odd integers exactly 1 greater than every integer divisible by 100?

Throughout this process, we must keep one thing in mind: Every sampling frame (or subframe) that we choose, and the final sample as well, must constitute an *unbiased representation* of the population for the purposes of our experiment. In any real-life experiment, the sample should not be too large or too small.

? Still Struggling

The optimum choice for the size of a sampling frame or sample presents a challenge. If you choose too large a sample for a statistical experiment, you'll have trouble collecting all the data because the process takes too many human-hours, or requires too much travel, or costs too much. If you choose too small a sample, it won't give you a fair representation of the population for the purposes of the experiment. As the sample gets smaller, the risk of its constituting a poor representation increases.

PROBLEM 5-1

Suppose that you want to describe the concept of a "number" to grammar-school students. In the process of narrowing down sets of numbers such as those described above into sampling frames in an attempt to make the idea of a number clear to a child, name a few possible assets, and a few limitations.

SOLUTION

Think back to your early years, say second or third grade. You had a good idea, even then, what constitutes a "whole number," didn't you? The concept of "whole number" might make a good sampling frame when talking about the characteristics of a number to a third-grader. By sixth grade, you knew about fractions, and therefore about rational numbers; the set of whole numbers would not have been a large enough sampling frame to satisfy you at age 11 or 12. But try talking about irrational numbers to a third grader! You won't get far. Nevertheless, a senior in high school would (we hope) know all about the real numbers and various subcategories of numbers within it. Restricting the sampling frame to the rational numbers would leave an 18-year-old unsatisfied. Beyond the real numbers lie the realms of complex numbers, vectors, quaternions, tensors, and transfinite numbers. Some people go through their whole lives without knowing much, if anything, about these esoteric quantities.

PROBLEM 5-2

Imagine that you want to quantify the effect (if any) that cigarette smoking has on people's blood pressure. You conduct the experiment on a world-wide basis, for all races of people, female and male. You interview people and ask them how much they smoke, and you measure their blood pressure levels. The population for your experiment comprises the set of all people in the world. Obviously you can't carry out the experiment for this entire population! Suppose, therefore, that you interview 100 people from each country in the world. The resulting group of people constitutes the sampling frame. Name some potential flaws with this scheme. Pose the issues as questions.

SOLUTION

Following are some questions that you'd have to answer or resolve before you could have confidence in the accuracy of the above-described experiment.

- How do you account for the fact that some countries have far more people than others?
- How do you account for the fact that the genetic profiles of the people in various countries differ?
- How do you account for the fact that people smoke more in some countries than in others?
- How do you account for the fact that the average age of the people in various countries differs, in light of the fact that scientists have found that age and blood pressure correlate?

- **How do you account for differing nutritional quality in various countries, a factor known to affect blood pressure?**
- **How do you account for differences in environmental pollutants, a factor that some scientists believe influences blood pressure levels?**
- **Does a set of 100 people in each officially recognized country constitute a large enough sampling frame?**

"Random" Sampling

When we want to analyze something in a large population and get an unbiased cross-section of that population, we can perform so-called *"random" sampling*. While some doubt exists as to whether we, or any machine, can generate a *truly random* set of numbers, we can get close to the theoretical ideal by making use of sets, lists, or tables of *pseudorandom numbers*. These sets, lists, and tables contain digits from the set {0, 1, 2, 3, 4, 5, 6, 7, 8, 9} chosen by machines programmed to output them with the least bias that technology allows. We'll take a closer look at the concept of randomness in Chap. 7.

A "Random" Sampling Frame

Consider the set T of all possible telephone numbers in the United States of America (USA) in which the last two digits equal 8 and 5, in that order. In theory, we can express any element t of the set T in the following format:

$$t = abc\text{-}def\text{-}gh85$$

where each letter a through h represents a digit from the set {0, 1, 2, 3, 4, 5, 6, 7, 8, 9}. The dashes do not represent minus signs; we include them to separate number groups, the standard convention for USA telephone designators.

If you're familiar with the format of USA phone numbers, you'll know that the first three digits a, b, and c together form the *area code*, and the next three digits d, e, and f represent the *dialing prefix*. Some values generated this way fail to constitute valid USA phone numbers. For example, the format above implies that 1000 different area codes exist, but there aren't that many area codes in the set of valid USA phone numbers. If a "randomly" chosen phone number isn't a valid USA phone number, let's reject it. Then we can generate pseudorandom digits for each of value a through h in the following generalized number:

$$a,bcd,efg,h85$$

where *h* represents the *hundreds digit*, *g* represents the *thousands digit*, *f* represents the *ten-thousands digit*, and so on up to *a*, which represents the *thousand-millions* (or, in the USA, the *billions*) digit. We can pick out eight-digit sequences from a pseudorandom-digit list, and plug them in repeatedly as values *a* through *h* to obtain 10-digit phone numbers ending in 85.

Smaller and Smaller

In the above scenario, we have a large sampling frame indeed! In reality, the number of elements is smaller than the maximum possible, because some of the telephone numbers derived by pseudorandom-number generation are not assigned to anybody.

Suppose that we conduct an experiment requiring us to get a "random" sampling of telephone numbers in the USA. By generating samples on the above basis, that is, picking out those that end in the digits 85, we're off to a good start. It's reasonable to think that this sampling frame constitutes an unbiased cross section of all the phone numbers in the USA. But we'll want to use a smaller sampling frame in order to conduct our experiment within the amount of time we have available.

What will happen if we go through a list of all the valid area codes in the USA, and throw out sequences *abc* that don't represent valid area codes? This process still leaves us with a sampling frame larger than the set of all assigned numbers in the USA. How about allowing any area code, valid or not, but insisting that the number *ab,cde,fgh* be divisible by 7 without a remainder? Within the set of all such possible numbers, we can find the set of all actual 10-digit phone numbers that end in 85 and whose preceding numeral is divisible by 7 (Fig. 5-2). By "actual," we mean any number that produces a connection when dialed. Imagine that we decide to use that set, portrayed at the center of the group of nested sampling frames in Fig. 5-2, as our final sampling frame: the one we'll work with to carry out our experiment.

Replacement

Just after we sample an element of a set, we can leave that element in the set, or else we can remove it. If we leave the element in the set so that we can sample it again, we perform *sampling with replacement*. If we remove the element so that we can't sample it again, we perform *sampling without replacement*.

When we do sampling without replacement starting with a finite sample set, we'll eventually exhaust the set. Table 5-1 shows how this process takes place if the initial sample contains 10 elements, in this case the first 10 letters of the English alphabet. This situation would presumably occur on a much larger scale in the hypothetical telephone-number experiment described above. The experimenters

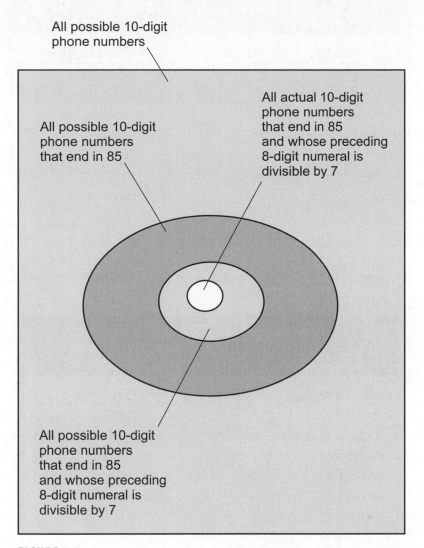

All possible 10-digit
phone numbers

All actual 10-digit
phone numbers
that end in 85
and whose preceding
8-digit numeral is
divisible by 7

All possible 10-digit
phone numbers
that end in 85

All possible 10-digit
phone numbers
that end in 85
and whose preceding
8-digit numeral is
divisible by 7

FIGURE 5-2 · An example of "nested" sampling frames.

would not want to count any particular telephone number twice, because that
would bias the experiment in favor of the repeated numbers.

If we conduct sampling with replacement, the size of the sample set
remains constant, regardless of how many elements it contains at the beginning. Table 5-2 shows how this process works with the first 10 letters of the
English alphabet. Note that once we've replaced an element in a finite sample
set, we should expect to encounter that element again during the course of
the experiment, as we do in the scenario of Table 5-2.

TABLE 5-1 When we do sampling without replacement, the size of the sample set decreases and eventually declines to zero. (Read down for experiment progress.)

Before Sampling	Element	After Sampling
{abcdefghij}	f	{abcdeghij}
{abcdeghij}	d	{abceghij}
{abceghij}	c	{abeghij}
{abeghij}	g	{abehij}
{abehij}	a	{behij}
{behij}	h	{beij}
{beij}	i	{bej}
{bej}	j	{be}
{be}	e	{b}
{b}	b	∅

TABLE 5-2 When we do sampling with replacement, the size of the sample set stays the same, and we can take some samples repeatedly. (Read down for experiment progress.)

Before Sampling	Element	After Sampling
{abcdefghij}	f	{abcdefghij}
{abcdefghij}	d	{abcdefghij}
{abcdefghij}	c	{abcdefghij}
{abcdefghij}	d	{abcdefghij}
{abcdefghij}	a	{abcdefghij}
{abcdefghij}	h	{abcdefghij}
{abcdefghij}	i	{abcdefghij}
{abcdefghij}	i	{abcdefghij}
{abcdefghij}	e	{abcdefghij}
{abcdefghij}	b	{abcdefghij}
{abcdefghij}	a	{abcdefghij}
{abcdefghij}	d	{abcdefghij}
{abcdefghij}	e	{abcdefghij}
{abcdefghij}	h	{abcdefghij}
{abcdefghij}	f	{abcdefghij}
↓	↓	↓

If the sample set contains infinitely many elements—the set of all points on a geometric line segment, for example, or the set of all integers—the size of the sample set doesn't decrease, even if we do not conduct replacements. We can never exhaust an infinite set when we choose elements one at a time, no matter how many of them we test and then toss out. In fact, if we keep taking away (or, for that matter, adding) elements to an infinite set one at a time, the size of the set, in mathematical terms, remains constant no matter how long we keep at the task!

PROBLEM 5-3

Here's a problem that gets into "fringe mathematics." Consider the set of all rational numbers between 0 and 1 inclusive. Because this set contains infinitely many elements, you can do one-by-one sampling from it forever, whether or not you replace the elements. If you carry out the sampling "at random" and without replacement, a specific number will never turn up twice during the course of the experiment, no matter how long you continue. But suppose that you do sampling with replacement. Can any given number turn up twice that way?

SOLUTION

If you attempt to calculate the probability of any given number in an infinite set turning up a second time after you've already picked it once, you'll get "one divided by infinity," which you can write as $1/\infty$. It's quite tempting to imagine that $1/\infty = 0$, even though most pure mathematicians will warn you that such a quotient has no real meaning because "infinity" isn't a number. However, if you accept that little bit of "fringe mathematics," you can conclude that any specific number can never turn up again once you've selected it from an infinite set—even if you replace it after you've picked it.

Minimizing Error

When we conduct a statistical experiment in the physical world, errors always occur. We can minimize the problem by designing and executing our experiments with care. The most common sources of *experimental defect error* include:

- A sample that's too small
- A sample that involves bias (favoritism)

- Neglecting factors that can introduce bias
- Inappropriate replacement of elements
- Failure to replace elements when necessary
- Attempting to compensate for factors that lack real effect
- Sloppy measurement of quantities when sampling

Improper tallying of the results of a poll or election constitutes a good example of sloppy measurement. If people don't observe the results correctly even though the machines work perfectly, human error will result. An election is a *binary digital* process; a voter casts either a "yes" (logic 1) or "no" (logic 0) for each candidate. (If voter makes no choice of candidate for a given elective office, then we can count that behavior as a "no" vote for all the candidates in that elective office.)

Measurement error can occur because of limitations in analog hardware. Suppose that we want to determine the average current consumed by commercially manufactured 550-watt sodium-vapor lamps when supplied with 90 volts rather than the normal rated 120 volts. We need an *alternating-current* (AC) *ammeter* (current-measuring meter) in order to conduct such an experiment. If the ammeter has a defect, the results will turn out inaccurate. No ammeter gives perfect readings, and with an *analog display* (the "needle-and-scale" type), we must contend with the additional human-error problem of *visual interpolation*. We would never have to worry, of course, about the consequences of using an ammeter with a *numeric digital display* that yielded *more* precision than required—say, a device that could measure currents down to within the nearest millionth of an ampere! Figure 5-3 shows what an analog ammeter might read if placed in the utility line along with a high-wattage, 120-volt lamp operating at only 90 volts.

PROBLEM 5-4

Interpolate the reading of the ammeter in Fig. 5-3 to the nearest tenth of an ampere, and to the nearest thousandth of an ampere.

SOLUTION

Note that the divisions get closer together as the value increases (toward the right). Based on this characteristic, and using some common sense, we can estimate the meter reading as 3.6 amperes, accurate to the nearest tenth of an ampere. However, we can't visually interpolate the reading

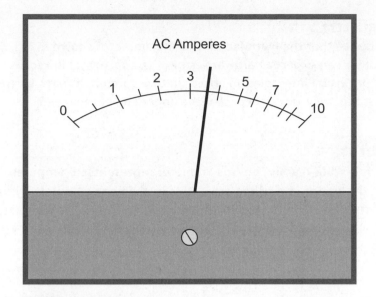

FIGURE 5-3 · A slight error occurs when we attempt to visually interpolate an analog meter reading.

much more precisely than the nearest tenth of an ampere in this region of
the scale, and we certainly can't interpolate it to the nearest thousandth of
an ampere at any point on the scale.

? Still Struggling

You ask, "How accurately can a human interpolate the readings of a meter such as
the one shown in Fig. 5-3?" We should expect an experienced engineer to say that
something like 1/5 (or 0.2) of an ampere constitutes the minimum visual-interpo-
lation error near the upper end of the scale, and something like 1/20 (or 0.05) of an
ampere represents the smallest visual error near the lower end of the scale. When
we want to "guesstimate" the accuracy of a visual interpolation, we should err on
the side of "conservatism." If we agree with this imaginary engineer, we would say
that the visual interpolation error equals roughly ±0.2 ampere near the upper end
of the scale, and roughly ±0.05 ampere near the lower end.

■ PROBLEM **5-5**

Suppose that the manufacturer of the ammeter shown in Fig. 5-3 tells us that we can expect a hardware error of up to ±10% of full scale, in addition to any visual interpolation errors that we commit. If that's true, what's the best accuracy we can expect at the upper end of the scale?

✔ SOLUTION

A full-scale reading on this ammeter represents 10 amperes. Therefore, at the upper end of the scale, this instrument could err by as much as ±(10% × 10), or ±1 ampere. Adding this to the visual interpolation error of ±0.2 ampere, we get a total measurement error of up to ±1.2 ampere at the upper end of the scale.

Estimation

In the real world, we can approximate characteristics such as the mean and standard deviation only on the basis of experimentation. We call this process *estimation*.

Estimating the Mean

Imagine that we have access to all of the 550-watt sodium-vapor lamps in the world; suppose that they number in the millions. Think about the set of all such lamps designed to operate at 120 volts AC, the standard household utility voltage in the USA. Suppose that we connect each and every one of these bulbs to a 90-volt AC source and measure the current that each bulb draws at this below-normal voltage. Suppose that we get current readings of around 3.6 amperes, as the meter in Fig. 5-3 shows. After rejecting obviously defective bulbs, each bulb we test produces a slightly different reading on the ammeter because of imprecision in the production process, just as the individual pieces in a bag full of jelly beans vary slightly in size.

We lack the time or resources to test millions of bulbs, so we select 1000 bulbs "at random" and test them. We obtain a high-precision digital AC ammeter that can resolve current readings down to the thousandth of an ampere, and we obtain a power supply that outputs something close enough to 90 volts so that we can consider that voltage as a mathematically

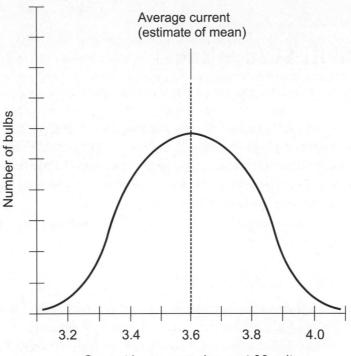

FIGURE 5-4 · Estimating the mean. The accuracy improves as we increase the size of the sample set.

exact value. We now have, in effect, an "ideal laboratory" with measurement equipment that does its job to perfection and eliminates human interpolation error. Variations in current readings therefore represent actual differences in the current drawn by different lamps. We plot the results of the experiment as a graph, smooth it out with a computer program designed to perform curve fitting, and end up with the normal distribution shown in Fig. 5-4.

Suppose that we find the arithmetic mean—the average—of all the current measurements and come up with 3.600 amperes, accurate to the nearest thousandth of an ampere. We have produced an *estimate of the mean* current drawn by all the 550-watt, 120-volt bulbs in the world when subjected to 90 volts. This estimate does not tell us the true mean, because we've tested only a small sample of the bulbs, not the whole population.

? **Still Struggling**

You wonder, "How can we determine the true mean current drawn by the foregoing type of bulb at 90 volts?" That's an excellent question. We can make our estimate more accurate by testing more bulbs (say 10,000 instead of 1000). We can save time and money by testing fewer bulbs (say 100 instead of 1000), but this shortcut would produce a less accurate estimate. In any event, we can *never* claim to know the true mean current drawn by this particular type of bulb at 90 volts unless we individually test every single 550-watt sodium-vapor lamp in the world.

Estimating the Standard Deviation

In the above imaginary situation, we can say that there exists a true distribution representing a graphical plot of the current drawn by each and every one of the 120-volt, 550-watt bulbs in the world when subjected to 90 volts. The fact that we lack the resources to find that distribution and exhibit it (we can't test every bulb in the world) does not mean that the distribution does not exist. As we increase the size of the sample, our rendition of Fig. 5-4 will approach the true distribution, and the average current (estimate of the mean current) will approach the true mean current.

True distributions can exist for populations so huge that we can consider them infinitely large "for all intents and purposes." For example, suppose that we want to estimate the mean power output of the stars in the known universe! All the scholars in the world, working together, can never hope to obtain an actual figure for a parameter such as this. The cosmos is too vast; the number of stars is too great. Nevertheless, the true mean power output of all the stars in the known universe constitutes a figure every bit as real as the fact that the stars are there.

The rules concerning estimation accuracy that apply to finite populations also apply to infinite populations. As the size of the sample set increases, the accuracy of the estimation improves. As the sample set becomes gigantic, the estimated value approaches the true value.

The mean current drawn at 90 volts does not, of course, constitute the only characteristic of the light-bulb distribution that we can estimate. We can also estimate the standard deviation as shown in Fig. 5-5. We derive the curve by

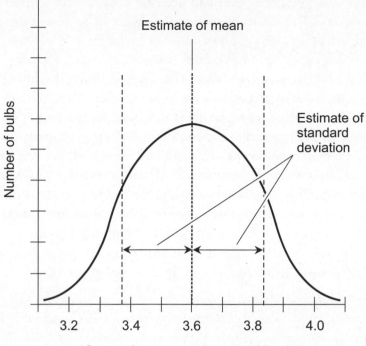

FIGURE 5-5 • Estimating the standard deviation. The accuracy improves as we increase the size of the sample set.

plotting the points, based on all 1000 individual tests, and then we "smooth out" the results with curve fitting. Once we have a graph of the curve, we can use a computer to calculate the standard deviation. From Fig. 5-5, it appears that the standard deviation σ equals approximately 0.23 amperes either side of the mean. If we test 10,000 bulbs, we'll get a more accurate estimate of σ. If we test only 100 bulbs, we'll get a less accurate estimate of σ.

Sampling Distributions

Imagine that, in the foregoing bulb-testing scenario, our sample consists of 1000 bulbs selected "at random," and we get the results illustrated by Figs. 5-4 and 5-5. What will happen if we repeat the experiment, again choosing a sample consisting of 1000 "randomly" selected bulbs? We won't get the same 1000 bulbs as we did the first time, so the results of the experiment will turn out a little differently.

If we do the experiment over and over, we'll work with a different set of bulbs every time. The results of each experiment will be almost the same, but not exactly identical. We'll observe a slight variation in the estimate of the mean from one experiment to another. Likewise, we'll see a small variation in the estimate of the standard deviation. The variation from experiment to experiment will increase if we make the sample size smaller (say 100 bulbs), and the variation will diminish if we make the sample size larger (say 10,000 bulbs).

Imagine that we repeat the experiment indefinitely, estimating the mean again and again. As we carry out this tedious task and plot the results, we obtain a distribution that tells us how the mean varies from sample to sample, obtaining a graph that looks something like Fig. 5-6. It's a normal distribution, but the values appear much more closely clustered around 3.600 amperes than those in Figs. 5-4 or 5-5. We might also plot a distribution that shows how the standard deviation varies from sample to sample. Again we'll get a normal distribution; its values are closely clustered around 0.23, as shown in Fig. 5-7.

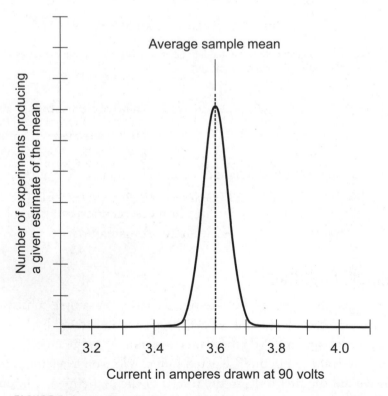

FIGURE 5-6 · Sampling distribution of means.

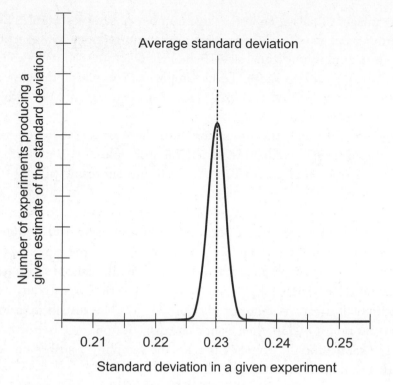

Average standard deviation

Number of experiments producing a given estimate of the standard deviation

Standard deviation in a given experiment

0.21   0.22   0.23   0.24   0.25

FIGURE 5-7 · Sampling distribution of standard deviations.

Figures 5-6 and 5-7 constitute examples of *sampling distributions*. Figure 5-6 shows a *sampling distribution of means*. Figure 5-7 illustrates a *sampling distribution of standard deviations*. If our experiments involved the testing of more than 1000 bulbs, these distributions would appear more centered (more sharply peaked curves), indicating less variability from experiment to experiment. If our experiments involved the testing of fewer than 1000 bulbs, the distributions would come out less centered (flatter curves), indicating greater variability from experiment to experiment.

The Central Limit Theorem

Imagine a population P in which some characteristic x can vary from element to element (or individual to individual). Suppose that P contains p elements, and p is a gigantic number. We plot the values of x on the horizontal axis of a graph, and we plot the number y of individuals with characteristic value x on the vertical axis. We thereby obtain a statistical distribution. The number of elements p is so great that we can render the distribution as a smooth curve.

Now imagine that we choose a large number of samples from P. Let's call that number k. Each sample represents a different "random" cross section of P, but all the samples have equal size. Each of the k samples contains n elements, where $n < p$. We find the mean of each sample and compile all these means into a set $\{\mu_1, \mu_2, \mu_3, ..., \mu_k\}$. We then plot these means on a graph. We end up with a sampling distribution of means.

We've gone through this discussion with the example involving the light bulbs, and now we're stating the underlying principle in general terms. We're repeating this concept because it leads to an important result called the *Central Limit Theorem*.

According to the first part of the Central Limit Theorem, the sampling distribution of means constitutes a normal distribution if the distribution for P is itself normal. If the distribution for P is not normal, then the sampling distribution of means approaches a normal distribution as the sample size n increases. Even if the distribution for P is highly *skewed* (asymmetrical), any sampling distribution of means is more nearly normal than the distribution for P. As things work out, if $n \geq 30$, then even if the distribution for P is grossly skewed and p is gigantic, for all practical purposes the sampling distribution of means constitutes a normal distribution.

The second part of the Central Limit Theorem concerns the standard deviation of the sampling distribution of means. Let σ represent the standard deviation of the distribution for some population P. Let n represent the size of the samples of P for which we determine a sampling distribution of means. We can calculate the standard deviation of the sampling distribution of means, more often called the *standard error of the mean (SE)*, using the formula

$$SE \approx \sigma/(n^{1/2})$$

This formula tells us that SE approximately equals the standard deviation of the distribution for P, divided by the square root of the number of elements in each sample. If the distribution for P is normal, or if $n \geq 30$, then we can consider the formula exact, and write it as

$$SE = \sigma/(n^{1/2})$$

TIP   *As the value of n increases, the value of SE decreases, reflecting the fact that large samples, in general, produce more accurate experimental results than small samples do. This general rule holds true up to about n = 30, beyond which we can't expect to achieve any better usable accuracy.*

Confidence Intervals

A distribution can provide generalized data about populations, but it doesn't tell us much about the individuals in the population. *Confidence intervals* give us a better clue as to what we can expect from elements taken "at random" from a population.

The Scenario

Imagine that we live in a remote scientific research outpost where the electric generators only produce 90 volts. The generators produce such low voltage (the standard utility voltage is around 120 volts) because they're old, inefficient, and too small. The government has drastically cut back funding for the outpost, so we can't afford to buy new generators. In short, we have too much energy demand and insufficient supply! (Sound familiar?)

We find it necessary to keep the outpost well-lit at night, regardless of whatever other sacrifices we have to make, so we need bright light bulbs. We have obtained data sheets for 550-watt sodium-vapor lamps designed for 120-volt circuits, and these sheets tell us how much current we can expect the lamps to draw at various voltages. Suppose that we've obtained the graph of Fig. 5-8,

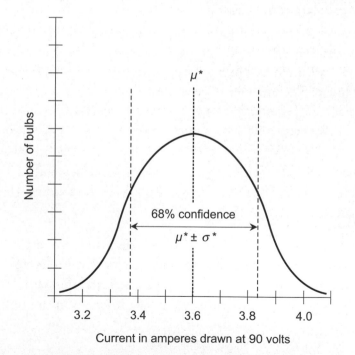

FIGURE 5-8 • The 68% confidence interval spans values in the range $\mu^* \pm \sigma^*$.

so we have a good idea of how much current each bulb will draw from our generators that produce 90 volts. We see that the estimate of the mean, μ^*, equals 3.600 amperes. Obviously, some bulbs draw a little more current than 3.600 amperes, and some draw a little less. A tiny proportion of the bulbs draw a lot more or less than 3.600 amperes of current.

If we pick a bulb "at random"—essentially what happens when anybody buys a single item from a large inventory—how much confidence can we have that our lamp will draw current within a certain range either side of 3.600 amperes?

68% Confidence Interval

Suppose that our data sheets indicate that the standard deviation of the distribution shown in Fig. 5-8 equals 0.230 amperes. According to the empirical rule, which we learned about in Chap. 3, we know that 68% of the elements in a sample have a parameter that falls within one standard deviation ($\pm\sigma$) of the mean μ for that parameter in a normal distribution. We don't know the actual standard deviation σ or the actual mean μ for the lamps in our situation, but we have estimates μ^* and σ^* that we can use to obtain a good approximation of a *confidence interval*.

In our situation, the parameter is the current drawn at 90 volts. Therefore, we can expect that 68% of the bulbs will draw current that falls in a range equal to the estimate of the mean plus-or-minus one standard deviation ($\mu^* \pm \sigma^*$). In Fig. 5-8, this range goes from a minimum of 3.370 amperes to a maximum of 3.830 amperes. It's a *68% confidence interval* because, if we select a single bulb, we can be 68% confident that it will draw current in the range between 3.370 and 3.830 amperes when we connect it to our antiquated 90-volt generator.

95% Confidence Interval

According to the empirical rule, 95% of the elements have a parameter that falls within two standard deviations of the mean for that parameter in a normal distribution. Again, we don't know the actual mean and standard deviation. We only have estimates of them, because the data is not based on tests of all the bulbs of this type that exist in the world. Nevertheless, we can use the estimates to get a good idea of the *95% confidence interval*.

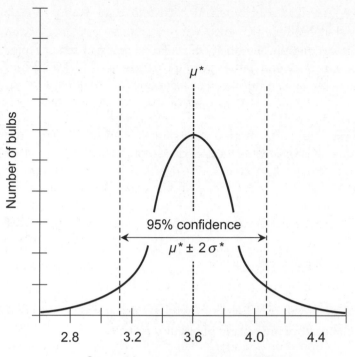

FIGURE 5-9 • The 95% confidence interval spans values in the range $\mu^* \pm 2\sigma^*$.

In our research-outpost scenario, we can reasonably expect that 95% of the bulbs will draw current that falls in a range equal to the estimate of the mean plus-or-minus two standard deviations ($\mu^* \pm 2\sigma^*$). In Fig. 5-9, this range goes from 3.140 amperes to 4.060 amperes.

You'll often read or hear 95% confidence intervals quoted in real-world situations. For example, someone says that "there's a 95% chance that Ms. X will survive her case of cancer for more than one year," or that "the probability is 95% that the center of Hurricane X will fail to strike Miami." If such confidence statements are based on historical data, we can regard them as reflections of truth. But the way we see them depends on the particulars of our situation. If you have an inoperable malignant tumor, or if you live in Miami and are watching a hurricane prowling the Bahamas, you may take some issue with the use of the word "confidence" when talking about your future. Statistical data can

> ### Are You a Skeptic?
>
> In some situations, the availability of statistical data can actually affect or alter the event for which you gather the data. Cancer and hurricanes cannot "care" about polls, but people can! If you hear, for example, that there's a "95% chance that Dr. J will beat Mr. H in the next local mayoral election," take it with skepticism. Inherent problems exist with this type of analysis, because people's reactions to the publication of predictive statistics can affect the actual event. If broadcast extensively by the local media, a statement suggesting that Dr. J has the election "already won" could cause overconfident supporters of Dr. J to stay home on election day, while those who favor Mr. H go to the polls in greater numbers than they would have if the data hadn't been broadcast. But who knows? The broadcast could have the opposite effect, causing supporters of Mr. H to stay home because they believe they'd waste their time going to an election if everyone expects their candidate to lose.

look a lot different to us when our own lives are at risk, as compared to when we sit in a laboratory measuring physical quantities.

99.7% Confidence Interval

The third phase of the empirical rule states that in a normal distribution, 99.7% of the elements in a sample have a parameter that falls within three standard deviations of the mean for that parameter. From this fact we can develop an estimate of the 99.7% *confidence interval*. In the outpost-lighting situation described earlier, we can expect that 99.7% of the bulbs will to draw current that falls in a range equal to the estimate of the mean plus-or-minus three standard deviations ($\mu^* \pm 3\sigma^*$). In Fig. 5-10, this range goes from 2.910 amperes to 4.290 amperes.

c % Confidence Interval

We can directly determine any confidence interval we want, within reason, from a distribution when we have good estimates of the mean and standard deviation (Fig. 5-11). The width of the confidence interval, specified as a percentage c, depends on the number of standard deviations x either side of the mean in a normal distribution. This relationship takes the form of a function of x versus c. When graphed for values of c ranging upward of 50%, the function of x versus c for a normal distribution looks like the curve shown in Fig. 5-12. The curve "blows up" at $c = 100\%$.

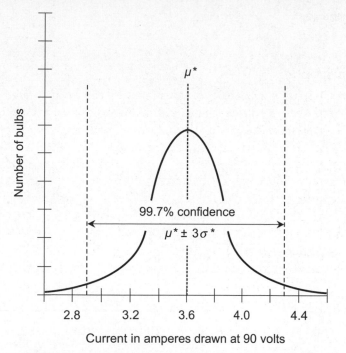

FIGURE 5-10 · The 99.7% confidence interval spans values in the range $\mu^* \pm 3\sigma^*$.

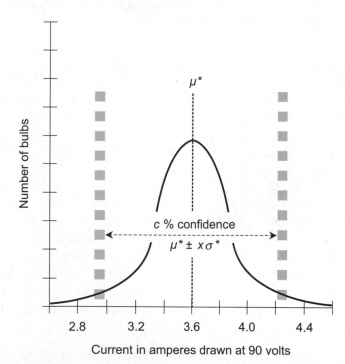

FIGURE 5-11 · The c% confidence interval spans values in the range $\mu^* \pm x\sigma^*$.

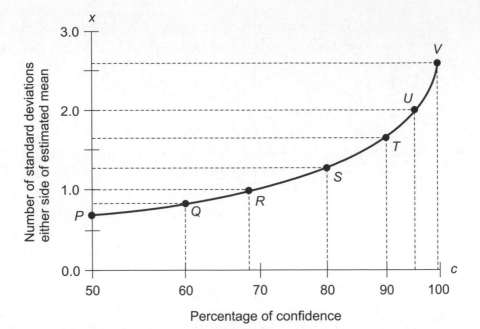

FIGURE 5-12 • Graph of *x* as a function of *c*. Points have the following approximate values, expressed as ordered pairs: *P* = (50%,0.67), *Q* = (60%,0.84), *R* = (68%,1.00), *S* = (80%,1.28), *T* = (90%,1.64), *U* = (95%,2.00), and *V* = (99%,2.58).

Don't Be Deceived!

We can never expect conclusions such as the foregoing to come out exact, for two important reasons.

- Unless we have a population small enough so that we can test every single element, we can only get estimates of the mean and standard deviation, never the actual values. We can overcome this problem, or at least minimize its effects, by using good experimental practice when we choose our sample frame and/or samples.
- When the estimate of the standard deviation σ^* represents a sizable fraction of the estimate of the mean μ^*, we get into trouble if we stray too many multiples of σ^* either side of μ^*. This problem gets worse as the parameter decreases. If we wander too far to the left (below μ^*), we get close to zero and risk stumbling into negative territory (e.g., "predicting" that we could find a light bulb that draws less than no current).

Calculations for large confidence intervals (many estimates of the standard deviation either side of the estimate of the mean) work *only* when the span of values constitutes a *small fraction* of the estimate of the mean. This situation holds true in the cases represented above and by Figs. 5-8 through 5-10. If the distribution were much flatter, or if we wanted a much greater degree of certainty, we could not specify such large confidence intervals without modifying the formulas.

PROBLEM 5-6

Suppose that you set up 100 archery targets, each target measuring exactly 1 meter in radius, and then let thousands of people shoot millions of arrows at these targets from 10 meters away. Each time a person shoots an arrow, you note and record the radius r at which the arrow hits the target, measured to the nearest millimeter from the exact center point P. If an arrow hits to the left of a vertical line L through P, you assign the radius a negative value; if an arrow hits to the right of L, you assign the radius a positive value (Fig. 5-13), obtaining a normal distribution in which the mean value μ of r equals 0.

Now imagine that you use a computer to create a graph of this distribution, plotting values of r on the horizontal axis and the number of shots for each

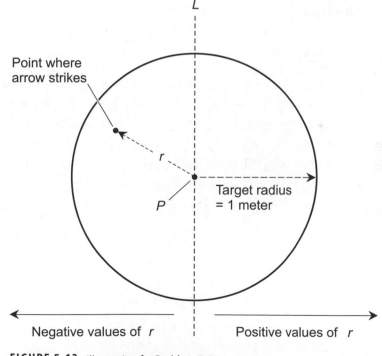

FIGURE 5-13 • Illustration for Problem 5-6.

value of r (to the nearest millimeter) on the vertical axis. You run a computer program to evaluate this curve and discover that the standard deviation, σ, of the distribution equals 150 millimeters. This number constitutes an actual value, not a mere estimate, because you've recorded the data for every single arrow that anyone ever shot during this experiment.

If you select a person "at random" from the people who have taken part at the experiment and have her shoot a single arrow at a target from 10 meters away, what's the radius of the 68% confidence interval? The 95% confidence interval? The 99.7% confidence interval? What do these figures actually tell you? Assume that no physical anomaly (such as a crosswind) exists that could complicate the experiment.

✔ SOLUTION

The radius of the 68% confidence interval equals σ, or 150 millimeters, meaning that you can have 68% confidence that the subject's shot will land within 150 millimeters of the center point P. The radius of the 95% confidence interval equals 2σ, or 300 millimeters, indicating that you can be 95% sure that the arrow will land within 300 mm of P. The 99.7% confidence interval equals 3σ, or 450 mm, so you can be 99.7% sure that the arrow will land within 450 mm of P.

PROBLEM 5-7

Draw a graph of the distribution resulting from the experiment described in Problem 5-6, showing the 68%, 95%, and 99.7% confidence intervals.

✔ SOLUTION

Figure 5-14 illustrates the curve, with the radius (to the nearest millimeter) along the horizontal axis and the number of shots along the vertical axis.

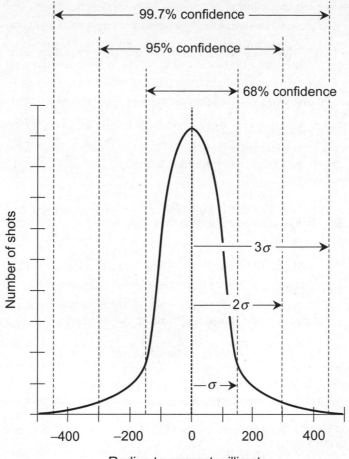

FIGURE 5-14 • Illustration for Problem 5-7.

QUIZ

Refer to the text in this chapter if necessary. A good score is 8 correct. Answers are in the back of the book.

1. When we do sampling with replacement in a finite set, the number of elements in the set
   A. increases one element at a time.
   B. decreases one element at a time.
   C. decreases more than one element at a time.
   D. never changes.

2. As we reduce the width of the confidence interval in a normal distribution, the probability that an element selected "at random" will lie within that interval
   A. decreases.
   B. does not change.
   C. increases.
   D. approaches 50%.

3. As we reduce the size of a sampling frame compared to the size of a population, what should we expect?
   A. It will constitute a better representation of the population.
   B. It will become increasingly "random."
   C. We will have to expend more time, energy, and money to evaluate it.
   D. It will constitute a less useful representation of the population.

4. We can directly figure out any desired confidence interval for a statistical distribution on the basis of
   A. the mean and the mode.
   B. the mode and the median.
   C. the mean and the standard deviation.
   D. the mode alone.

5. When a statistician evaluates the data gathered by someone else (a biologist, for example), we call it
   A. independent data.
   B. pseudorandom data.
   C. direct data.
   D. None of the above

6. The 95% confidence interval in a normal distribution spans values within plus or minus
   A. four standard deviations of the estimate of the mean.
   B. three standard deviations of the estimate of the mean.
   C. two standard deviations of the estimate of the mean.
   D. one standard deviation of the estimate of the mean.

7. If we want to make sure that we get a reasonably accurate confidence-interval determination in a normal distribution, we must make sure that the span of values

    A. remains much smaller than the estimate of the mean.
    B. exceeds plus-or-minus one standard deviation of the estimate of the mean.
    C. constitutes a whole-number fraction of the estimate of the mean.
    D. constitutes a whole-number multiple of the estimate of the mean.

8. Suppose that we want to determine the percentage of people in the town of Hoodooburg, total population 100,000, who exercise more than 30 minutes a day (on the average). Which of the following samples should we expect would have the *most* bias?

    A. All of the people in Hoodooburg who belong to the swimming team
    B. All of the people in Hoodooburg who live on odd-numbered streets
    C. All of the people in Hoodooburg who were born in the month of November
    D. All of the people in Hoodooburg who have lived there for their entire lives

9. Which of the following *practical* methods would result in the *worst* estimate of the standard deviation for a characteristic of a population having 1,000,000,000,000 elements?

    A. Test 5000 elements chosen "at random."
    B. Test all of the elements in the population.
    C. Test the elements in the largest sample set that we can manage.
    D. Test the first 100 elements.

10. Consider a normal distribution that describes a certain characteristic K of the people who live in the town of Hoodooburg. According to the Central Limit Theorem, we know that the sampling distribution of means for K is

    A. flat.
    B. normal.
    C. uniform.
    D. skewed.

chapter 6

Hypotheses, Prediction, and Regression

In this chapter, we'll look at some of the techniques statisticians use to conduct experiments, interpret data, and draw conclusions.

CHAPTER OBJECTIVES

In this chapter, you will

- Formulate and test hypotheses
- Make null and alternative hypotheses
- Generate inferences and forecasts
- Quantify and graph regression between variables
- Learn the law of least squares
- Plot least-squares lines

Assumptions and Testing

We formulate a *hypothesis* whenever we make a supposition or assumption based on experience, reason, or educated opinion. We can formulate a hypothesis merely to see what will happen (or what we expect to happen) under certain circumstances. Sometimes, the truth or falsity of a hypothesis can actually affect or alter the outcome of a statistical experiment.

The Hypothesis as a Forecast

A hypothesis can take the form of a prediction or forecast. Sometimes such a prediction will turn out correct; sometimes it won't. Once in awhile, we'll make a hypothesis and never know whether or not the truth bore it out!

When statisticians collect data in a real-world experiment, some error always occurs. The error can result from instrument hardware imperfections, limitations of the human senses in reading instrument displays, and even plain carelessness. But in some experiments or situations, a more sinister source of potential error can "rear its head": vital data goes missing. Data can get left out because the researcher can't obtain it, or because someone omitted it with the intent of influencing the results of the experiment. In such a situation, we might have to make an "educated guess" as to the content of the missing data. Such a "guess" can take the form of a hypothesis.

Imagine a major hurricane named *Emma* churning in the North Atlantic. Suppose that you live in Wilmington, Delaware, and your house sits on the shore of the Delaware River. Does the hurricane pose a threat to you? If so, when should you expect the greatest danger to exist? If the hurricane strikes at your location, how serious will the situation get? The answers to these questions constitute dependent variables because they depend on several factors. A meteorologist can observe some of the factors and derive accurate forecasts based on them. Some of the factors will prove difficult to observe or forecast, so the meteorologist can only approximate them.

In a situation like this, a scientist can formulate a graph that shows the probability that the hurricane will follow a path between two geographical limits. Figure 6-1 shows an example of such a plot for our hypothetical Hurricane Emma. Percentage numbers indicate the probability that the storm will go somewhere—*anywhere*—between two limiting paths. The values increase as the limiting paths get farther away, in either direction, from the predicted path, which lies midway between the two centermost dashed-and-dotted lines, bearing the label 25%.

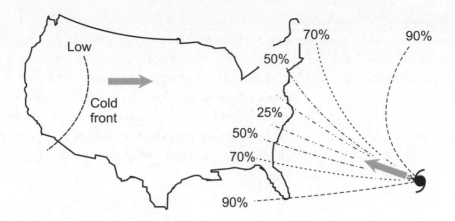

FIGURE 6-1 · When we want to predict where a hurricane will go, we must formulate hypotheses in addition to gathering the physical data.

Multiple Hypotheses

The weather experts use several *models* ("artificial worlds" usually generated with the help of computers) to make path predictions for Hurricane Emma. Each model uses data, obtained from instrument readings and satellite imagery, and processes the data using specialized software. The various models "imagine the world" in different ways and according to different parameters, so they don't necessarily all agree. In addition to the computer programs, the hurricane experts use historical data, and also some of their own intuition, to come up with an official storm path forecast and an official storm intensity forecast for the next 24 hours, 2 days, 3 days, and 5 days.

Imagine that, in our hypothetical scenario, all the computer models agree on one thing: Emma, which has evolved into a category-five hurricane (the most violent possible, with winds of 156 miles an hour or more), will remain at this level of intensity for the next several days, unusual for any tropical storm! The weather experts also agree that if Emma strikes land, people in its path will remember it for the rest of their lives. If the hurricane passes over Wilmington, the Delaware River will experience massive tidal flooding. You, who live on the riverfront, do not want to find yourself in the storm's way. You also want to take every reasonable precaution to protect your property from damage in case the river rises.

Along with the data, scientists have hypotheses to work with. Imagine that a large cyclonic weather system (called a *low*) exists, containing a long trailing cold front, moving from west to east across the USA. A "continental low" like this can pull a hurricane into or around itself. The hurricane tends to move toward and then follow the low, as if the hurricane were a rolling ball and the front were a trough. (That's where the expression "trough" comes from in weather jargon.) But this phenomenon only happens when a hurricane wanders close enough to a low to get caught up in the low's counterclockwise wind circulation. If this effect occurs with Emma, the hurricane will likely veer away from the coast, or else make landfall further north than expected. Will the low, currently over the western United States, affect Emma? No one knows yet, so we formulate two distinct hypotheses:

- The low crossing North America will move fast and will interact with Emma before the hurricane reaches land, causing the hurricane to follow a more northerly path than that implied by Fig. 6-1
- The low crossing North America will stall or dissipate, or will move slowly, and Emma will follow a path near, or to the south of, the one implied by Fig. 6-1

Suppose that we enter the first hypothesis into our various computer models along with the known data. In effect, we treat the hypothesis as a fact. The computer models create a forecast for us, producing the map of Fig. 6-2A. Then we enter the second hypothesis into the computer programs and let the machine grind out its forecast, producing the map of Fig. 6-2B.

If we want to get into more detail, we can consider the hypothesis as an independent variable, assign various forward speeds to the low-pressure system and the associated front, enter several forward-speed values into the computer models, and let the computer generate separate forecast maps for each input value.

Null Hypothesis

Suppose that we observe the speed of the continental low and its cold front as they cross the USA from west to east. We use the official weather forecasts to estimate the low's speed over the next few days. We input this data into a hurricane-forecasting program, which tells us that the *mean path* for Emma will take the hurricane across the mid-Atlantic coast of the USA. The computer generates the graphic of Fig. 6-3.

We now have a hypothesis concerning the future path of Hurricane Emma. The actual outcome, of course, remains independent of human control. If we decide to put the hypothesis illustrated by Fig. 6-3 to the test, we call it the *null hypothesis* and symbolize it as H_0 (read "H-null" or "H-nought").

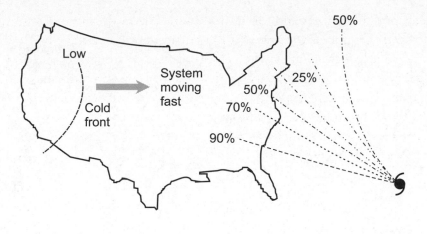

A

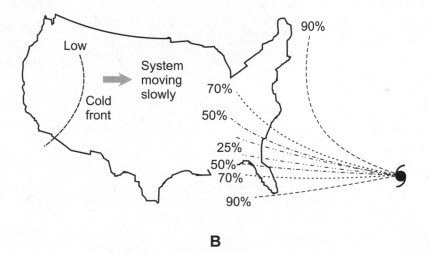

B

FIGURE 6-2 • Two hypotheses. At A, path probabilities for a hurricane with a fast-moving weather system on the continent. At B, path probabilities for a hurricane with a slow-moving weather system on the continent.

Alternative Hypotheses

People who live in or near the predicted path of Emma, as shown in Fig. 6-3, hope that H_0 will turn out wrong. The *alternative hypothesis*, symbolized H_1, constitutes the proposition that Emma will *not* follow the path near the one shown in Fig. 6-3. If someone asserts that Emma will go either north or south

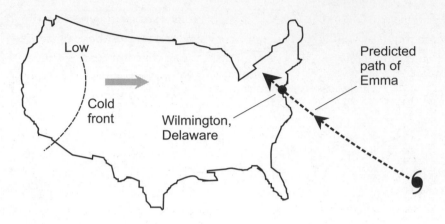

FIGURE 6-3 · A null hypothesis constitutes a claim or prediction, in this case a forecast path for Hurricane Emma.

of the path assumed by H_0, then that person proposes a *two-sided alternative*. If someone else claims that Emma will travel north of the path assumed by H_0, then that person proposes a *one-sided alternative*. If yet another person proposes that Emma will go south of the path assumed by H_0, that person also proposes a one-sided alternative.

In a situation of this sort, we should not experience too much surprise if numerous people come out and propose hypotheses such as the following:

- Emma will hit Washington, DC
- Emma will hit New York City
- Emma will hit between New York City and Boston, Massachusetts
- Emma will hit Houston, Texas
- Emma will hit San Francisco, California
- Emma will hit Juneau, Alaska

The number of possible hypothesis in this scenario is limited only by the number of different opinions that people can concoct. Some hypotheses, of course, have greater plausibility than others. We can take seriously the notion that Emma will strike somewhere between New York and Boston. Most people would reject the hypothesis that Emma will hit Houston (although stranger things have happened with hurricanes; Emma could take a sharp southward turn, go around the Florida peninsula, and get into the Gulf of Mexico). Any competent meteorologist would dismiss straightaway the hypotheses involving

San Francisco or Juneau, both of which lie on the Pacific coast of the USA. Emma might as well go to the moon as strike at either of those places!

Some people, noting historical data showing that hurricanes almost never strike the mid-Atlantic coast of the USA in a manner such as that given by the computer model and illustrated in Fig. 6-3, claim that the storm will stay further south. Other people think that Emma will travel north of the predicted path. We have good reason to believe either of these alternative hypotheses. In the past 100 years or so, the Carolinas and the Northeastern USA have taken direct hits from Atlantic hurricanes more often than has Delaware. Government agencies, academic institutions, corporations, and think tanks employ various computer programs designed to forecast the paths of hurricanes. Each program produces a slightly different mean path prediction for Emma, given identical input of data. Alternative hypotheses abound. The null hypothesis H_0 constitutes a lonely proposition, indeed.

? Still Struggling

How do we know whether or not a null hypothesis H_0 is correct? In order to find out, we must conduct an experiment; we must see the "real world" in action. In the case of Emma, that involves no active work on our part (other than preparing ourselves for the worst even as we hope for the best). We must wait and see what happens.

Friends of Vanilla

Let's forget about violent weather and forecasts of destruction, and instead imagine that we want to find out what proportion of ice-cream lovers in Canada prefer plain vanilla over all other flavors. Someone makes a claim that 25% of Canadian ice-cream connoisseurs go for vanilla, and 75% of that same group prefer some other flavor. We plan to test this hypothesis by conducting a massive survey. We have our null hypothesis, and we name it H_0.

Our basic alternative hypothesis, H_1, comprises the claim that H_0 is inaccurate. If an elderly woman says that the proportion of Canadian ice-cream devotees who prefer vanilla must exceed 25%, she asserts a one-sided alternative. If a young boy says "No sane person would like vanilla" (evidently expressing a belief that the proportion must be much lower than 25%), he also asserts a one-sided alternative. If a woman says that the proposition must be either larger or smaller than 25%, then she asserts a two-sided alternative. The experiment, of course, involves sheer labor: Do the survey and document the results.

Testing

In the "USA hurricane scenario," we can reasonably expect that we'll reject H_0 after the experiment has taken place. Even though Fig. 6-3 represents the mean path for Emma as determined by a computer program, the probability remains low that Emma will follow this precise path. If you find this confusing, you can think of it in terms of a double negative. The computers don't say that Emma will "probably" follow the path shown in Fig. 6-3. However, the computers do tell us that the displayed curve constitutes the *least unlikely individual path* according to the particular set of data and parameters that we have input to the program. (When we talk about the "probability" that something will or won't occur, we mean the degree to which the forecasters believe it, based on historical and computer data. We want to avoid the probability fallacy!)

Similarly, in the "Canada ice-cream scenario," the probability remains low that the proportion of vanilla aficionados among Canadian ice-cream connoisseurs equals 25% within the measurable margin of error. Even if we make this claim, we must expect that the experiment will yield results that differ from 25%. When we accept the null hypothesis H_0 in this case, we assert that all other exact proportion figures are less likely than 25%. (When we talk about the "probability" that something does or does not reflect reality, we mean the degree to which we believe it, based on experience, intuition, or plain guesswork. Again, we don't want to commit the probability fallacy.)

Whenever someone makes a prediction or states a claim, someone else will refute it. In part, this fact reflects human nature. But logic also plays a role. Computer programs for hurricane forecasting improve with each passing year. Methods of conducting statistical surveys about all subjects, including people's ice-cream flavor preferences, also get better as time passes.

If a group of meteorologists comes up with a new computer program, and that program tells them that Hurricane Emma will pass over New York City instead of Wilmington, then the output of that program constitutes evidence against H_0 in the "USA hurricane scenario." If someone produces the results of a survey showing that only 15% of British ice-cream lovers prefer plain vanilla flavor and only 10% of USA ice-cream lovers prefer it, we can consider that as reasonable evidence against H_0 in the "Canada ice-cream scenario." We call the gathering and presentation of data supporting or refuting a null hypothesis, and the conducting of experiments to figure out the true situation, *statistical testing* or *hypothesis testing*.

Forecasting

Let's revisit the "USA hurricane scenario." The predicted path for Hurricane Emma shown in Fig. 6-3 represents a hypothesis, not a fact. We'll learn the facts eventually, and meteorologists will draw a known path for Emma—because in a few days, the whole scenario will constitute nothing more than a memory!

The "Purple Line"

If you live in a hurricane-prone region, perhaps you've checked the Internet to get an idea of whether or not a certain storm threatens you. You look on the tracking/forecast map, and see that the experts have drawn a "purple line" (also known as the "line of doom") going over your town on the map! This line does not say that the forecasters think the storm will hit you for certain. It tells you that the "purple line" represents the *mean predicted path* based on computer models.

As time passes—that is, as the experiment plays itself out—you'll get a better idea of how much danger you face. If you're an unemotional, scientific person, you'll go to the Internet sites of the government weather agencies such as the National Hurricane Center, analyze the data for awhile, and then make whatever preparations you deem necessary. Perhaps you'll decide to take a short vacation to an inland town such as Nashville, Tennessee, and conduct your statistical analysis of Emma from there.

Confidence Intervals Revisited

Instead of drawing a single line on a map, indicating a predicted track for Hurricane Emma, you can draw path probability maps such as the ones in Figs. 6-1 and 6-2. These maps, in effect, show confidence intervals. As the storm draws closer to the mainland, the confidence intervals narrow. The meteorologists revise the forecasts. The "purple line"—the mean path of the storm—might shift on the tracking map. (Then again, maybe, it won't move at all.) Most hurricane Web sites have strike-probability maps that provide more information than the path-prediction maps do.

Probability Depends on Range

Imagine that a couple of days pass, Emma has moved closer to the mainland, and the weather experts still predict a mean path that takes the center of Emma over Wilmington. The probability lines appear more closely spaced on the maps than they were 2 days ago. We can generate a distribution curve that shows the relative danger at various points north and south of the predicted point of landfall (which actually lies on the New Jersey coast east of Wilmington). Figure 6-4A

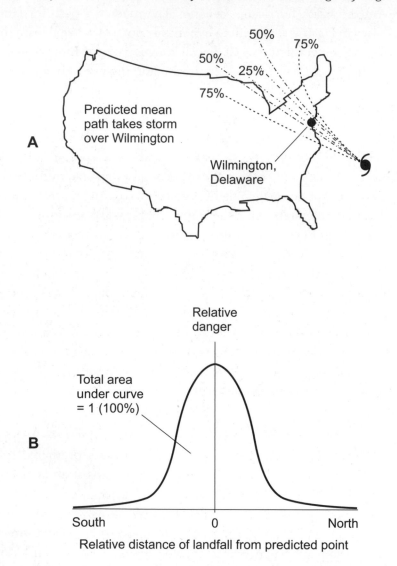

FIGURE 6-4 · At A, path probabilities for Hurricane Emma as it moves closer to the mainland. At B, relative plot of the strike danger as a function of the distance from the predicted point of landfall.

illustrates a path probability map, and Fig. 6-4B provides an example of a statistical distribution showing the relative danger posed by Emma at various distances to the north and south along the coastline from the predicted point of landfall. The vertical axis representing the landfall point (labeled 0 in Fig. 6-4B) does not depict the actual probability of strike. We can ascertain the strike probabilities only within various ranges—lengths of coastline—north and/or south of the predicted point of landfall as shown in Fig. 6-5A and B.

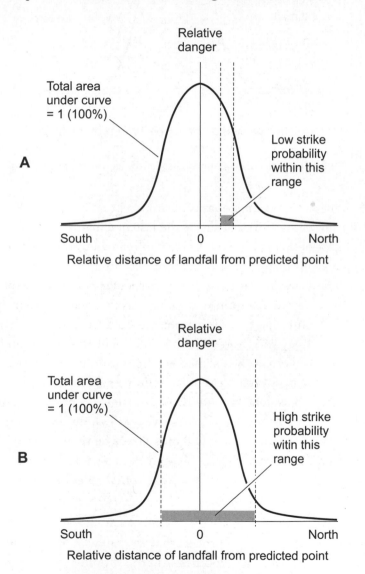

FIGURE 6-5 · At A, Hurricane Emma is unlikely to strike anywhere within narrow span of coast. At B, Hurricane Emma is likely to strike somewhere within a wide span of coast.

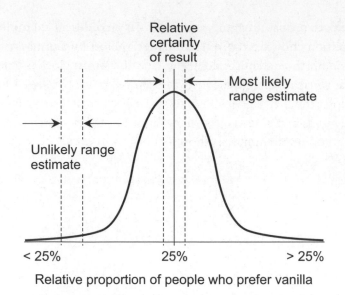

FIGURE 6-6 · Distribution for the "Canada ice-cream scenario."

In the "Canada ice-cream scenario," a similar situation exists. We can draw a distribution curve (Fig. 6-6) that shows the taste inclinations of people, based on the H_0 that 25% of them like plain vanilla ice cream better than any other flavor. Given any range or "margin of error" that represents a fixed number of percentage points wide (say ±2% either side of a particular point), H_0 asserts that the area under the curve and within that range will be greatest when the range is centered at 25%. Imagine that H_0 happens to hold true. In that case, someone who says "Our survey will show that 23% to 27% of the people prefer vanilla" will more likely be correct than someone who says "Our survey will show that 12% to 16% of the people prefer vanilla." In more general terms, let P represent some percentage between 0% and 100%, and let x represent a value much smaller than P. If someone says "Our survey will show that $P\% \pm x\%$ of the people prefer vanilla," that person will most likely be right if $P\% = 25\%$. That's where the distribution curve of Fig. 6-6 comes to a peak, and that's where the area under the curve, given any constant horizontal range, is the greatest.

Inference

The term *inference* refers to any process that we follow in order to draw conclusions on the basis of data and hypotheses. In the simplest sense, inference comprises the application of common sense. In statistics, inference requires the use of rigorous logic in specialized practical ways.

We've seen two tools that we can employ for statistical inference, both of which give us numerical output. But ultimately, we must make a subjective judgment as to whether we should come to any particular conclusion based on such data. Sometimes we can make such a judgment easily. In other cases, we'll find it difficult to make the judgment. Sometimes, we can make inferences and draw conclusions with a "cool head," because nothing important depends on our decision. In other cases, emotional or "life-and-death" factors can adversely affect our ability to make a logical judgment; then we run the risk of making an inference when we should not, or else failing to make an inference when we should.

Consider again the "USA hurricane scenario." If you live on the oceanfront and a hurricane looms, what should you do? Board up the windows? Go to a shelter? Find a friend who lives in a house with more solid construction than your house has? Get in your car and flee? Statistics can help you decide what to do, but no numbers exist that can mathematically define an optimal course of action. No computer can tell you the best thing to do. You have to make that decision for yourself.

In the "Canada ice-cream scenario," suppose that we conduct a survey by interviewing 12 people. Three of them (that's 25%) say that they prefer vanilla. Does it follow that H_0, our null hypothesis, is correct? Most people would say no, because 12 people doesn't constitute a big enough sample. But if we interview 12,000 people (taking care that the ages, ethnic backgrounds, and other factors give us an unbiased cross section of the Canadian population) and 2952 of them say they prefer vanilla, we can reasonably infer that H_0 is valid, because 2952 equals 24.6% of 12,000, and that's close to 25%. If 1692 people say they prefer vanilla, we can infer that H_0 does not hold true, because 1692 equals only 14.1% of 12,000, and that's nowhere near 25%.

? Still Struggling

You ask, "How large a sample should we obtain in order to take the results of our survey seriously?" That's a subjective decision. A dozen people isn't enough, and 12,000 is plenty; few people will dispute those notions. But what about 120 people? Or 240? Or 480? In a situation like this, we should gather as large a sample as we can, consistent with available time and resources, while making sure that the sample is unbiased.

PROBLEM 6-1

Imagine that you're a man who lives in a town of 1,000,000 people. Recently, you've seen a lot of women smoking cigarettes. You start to suspect that more female smokers than male smokers exist in your town. You discuss your observations and suspicions with a friend.

Your friend says, "Your fears have no basis in fact. The ratio of female smokers to male smokers in this town is 1:1."

You say, "Do you mean to tell me that the number of women smokers *equals* the number of men smokers?"

Your friend says, "Yes, or in any case, the numbers are very nearly equal."

You counter, "I see far more women smokers than men smokers every evening. It seems that almost every woman has a cigarette in her mouth."

Your friend retorts: "You see that only because you spend a lot of time at night clubs, where the number of women smokers is *out of proportion* to the number of women smokers in the general population."

You and your friend decide to conduct an experiment to resolve the confusion. You intend to prove that more female smokers than male smokers live in your town. Your friend offers the hypothesis that the number of male smokers equals that of female smokers. What's a good null hypothesis here? What's the accompanying alternative hypothesis? How might you conduct a test to find out who is right?

SOLUTION

A reasonable null hypothesis, which your friend proposes, says the notion that the ratio of female to male smokers in your town equals 1:1. Then the alternative hypothesis, which you propose, says that considerably more female smokers than male smokers exist. To find out who's right and who's wrong, you must choose an unbiased sample of the population of your town. The sample must comprise an equal number of men and women, and it should be as large as possible. You'll have to ask all the subjects whether or not they smoke, and then assume that they've told you the truth.

PROBLEM 6-2

Now imagine, for the sake of argument, that H_0 does in fact hold true in the above-described scenario. (You don't know it and your friend doesn't know it, because you haven't conducted the survey yet.) You're about to conduct an experiment by taking a supposedly unbiased survey consisting of 100 people: 50 men and 50 women. Draw a simple graph showing the relative probabilities of the null hypothesis being verified, versus either one-sided alternative.

✔ SOLUTION

The curve portrays a normal distribution as shown in Fig. 6-7. Of all the possible outcomes, the most likely represents an even split, in which the same number of women as men say that they smoke. This *exact* result (say, 33:33 or 24:24) isn't certain or even likely, but it's the *least unlikely* of all the possible outcomes. You can reasonably expect a result of, say, 21 women asserting that they smoke while 22 men say that they smoke. But you should not expect that the survey will reveal that 40 women smoke while only 10 men do, or that only one woman smokes while 45 men do.

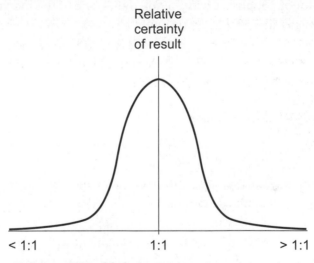

FIGURE 6-7 • Illustration for Problem 6-2.

PROBLEM 6-3

Name three different possible outcomes of the experiment described above, all of which apparently verify the null hypothesis.

☑ **SOLUTION**

Recall that you've surveyed 50 men and 50 women. If 20 men say that they smoke and 21 women say that they smoke, this suggests the null hypothesis is reasonable. The same goes for ratios such as 12:14 or 17:15.

◻ **PROBLEM 6-4**

Name two outcomes of the experiment described above, in which the null hypothesis is apparently verified, but in which the results should be highly suspect.

☑ **SOLUTION**

If all of the men and all of the women say they smoke, you ought to suspect that a lot of people are lying. Similarly, if none of the men and none of the women say they smoke, you should also expect deception. Even ratios of 3:4 or 47:44 should cause us to believe that something's wrong. (Results such as this would suggest that we conduct other experiments concerning the character of the people in this town!)

Regression

Regression allows us to quantify the extent of the relationship between two variables. We can employ regression (also known as *regression testing*) in an attempt to predict things, but we must use caution. The existence of a correlation between variables does not always logically imply that a cause-and-effect link exists between them.

Paired Data

Imagine two cities, one named Happyton and the other named Blissville, located far apart on the same continent in the Northern Hemisphere. The prevailing winds and ocean currents produce greatly different temperature and rainfall patterns throughout the year in these two cities. Suppose that we plan to move from Happyton to Blissville, and people have told us that Happyton "has soggy summers and dry winters," while in Blissville we should expect that "the summers die of thirst and the winters drown." People have also warned us that the average temperature varies drastically between summer and winter in Happyton (as we know, having dwelt there for years), but varies much less between summer and winter in Blissville.

We go to the Internet and begin to gather data about the two towns. We find a collection of tables showing the average monthly temperature in degrees Celsius and the average monthly rainfall in centimeters for many places throughout the world. Happyton and Blissville appear among the cities shown in the tables. Table 6-1A shows the average monthly temperature and rainfall for Happyton as gathered over the past 100 years. Table 6-1B shows the average monthly temperature and rainfall for Blissville over the same period. The data we have found is called *paired data*, because it portrays two variable quantities, temperature and rainfall, side by side.

We can get an idea of the summer and winter weather in both towns by scrutinizing the tables. But we can get a more visual-friendly portrayal by making use of bar graph.

TABLE 6-1A Average monthly temperature and rainfall for Happyton.		
Month	Average Temperature, Degrees Celsius	Average Rainfall, Centimeters
January	2.1	0.4
February	3.0	0.2
March	9.2	0.3
April	15.2	2.8
May	20.4	4.2
June	24.9	6.3
July	28.9	7.3
August	27.7	8.5
September	25.0	7.7
October	18.8	3.6
November	10.6	1.7
December	5.3	0.5

TABLE 6-1B   Average monthly temperature and rainfall for Blissville.

Month	Average Temperature, Degrees Celsius	Average Rainfall, Centimeters
January	10.5	8.1
February	12.2	8.9
March	14.4	6.8
April	15.7	4.2
May	20.5	1.6
June	22.5	0.4
July	23.6	0.2
August	23.7	0.3
September	20.7	0.7
October	19.6	2.4
November	16.7	3.4
December	12.5	5.6

Paired Bar Graphs

Let's graphically compare the average monthly temperature and the average monthly rainfall for Happyton. Figure 6-8A is a *paired bar graph* showing the average monthly temperature and rainfall there. We base this graph on the data from Table 6-1A. The horizontal axis has 12 intervals, each one showing a month of the year. Time constitutes the independent variable. The left-hand vertical scale portrays the average monthly temperatures, and the right-hand vertical scale portrays the average monthly rainfall amounts. Both of these are dependent variables—functions of the time of year. We portray the average monthly temperatures as light gray bars and the average monthly rainfall amounts as dark gray bars. We can see from this data that the temperature and rainfall both follow annual patterns. In general, the warmer months have more precipitation than the cooler months in Happyton.

Now let's make a similar comparison for Blissville. Figure 6-8B is a paired bar graph showing the average monthly temperature and rainfall there, based

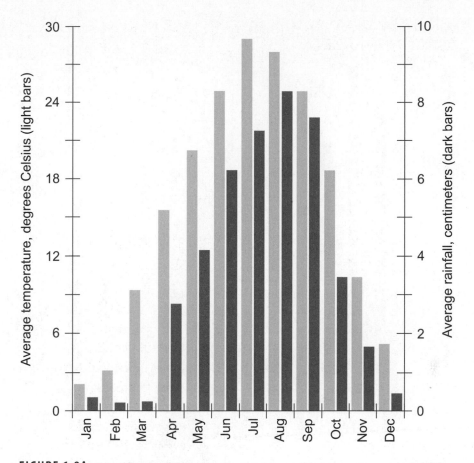

FIGURE 6-8A • Paired bar graph showing the average monthly temperature and rainfall for the hypothetical city of Happyton.

on the data from Table 6-1B. From this data, we can see that the temperature varies less between winter and summer in Blissville than it does in Happyton. But the rainfall profile for Blissville, as a function of the time of year, differs drastically from rainfall profile for Happyton. The two towns are exact opposites in this respect! The winters in Blissville, especially the months of January and February, have a lot of precipitation, while the summers, particularly June, July, and August, get almost none. The contrast in general climate between Happyton and Blissville is striking. We can infer this information if we scrutinize the tabular data, but we can see it instantly when we glance at the paired bar graphs.

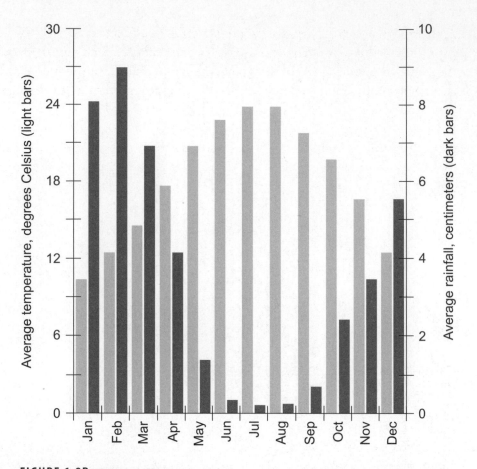

FIGURE 6-8B · Paired bar graph showing the average monthly temperature and rainfall for the hypothetical city of Blissville.

Scatter Plots

When we examine Fig. 6-8A, it appears that a relationship exists between temperature and precipitation for the town of Happyton. In general, as the temperature increases, so does the amount of rain. Evidently, a relationship also exists between temperature and rainfall in Blissville (Fig. 6-8B), but it goes in the opposite sense: as the temperature increases, the rainfall decreases. How strong are these relationships? We can draw *scatter plots* to find out.

Figure 6-9A shows the average monthly rainfall as a function of the average monthly temperature for Happyton. We plot one point for each month, based on the data from Table 6-1A. In this graph, the temperature, not the time of the year, constitutes the independent variable. We can see a pattern in the arrangement of the points. The correlation between temperature and rainfall is positive for Happyton. It's fairly strong, but not extremely so. If no correlation existed whatsoever (i.e., if it were 0), we'd see points scattered all over the graph. But if the correlation were perfect (either +1 or −1), all the points would lie along a straight line.

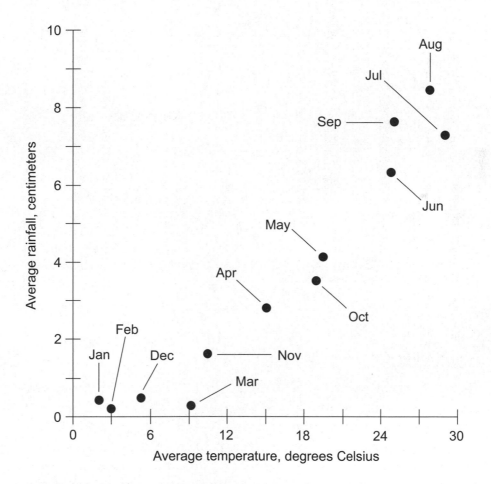

FIGURE 6-9A • Scatter plot of points showing the average monthly temperature versus the average monthly rainfall for Happyton.

Figure 6-9B shows a plot of the average monthly rainfall as a function of the average monthly temperature for Blissville. Again, we plot one point for each month, this time based on the data from Table 6-1B. As in Fig. 6-9A, temperature is the independent variable. We can see a pattern in the arrangement of points here, too. In this case the correlation is negative instead of positive. It's a fairly strong correlation, perhaps a little stronger than the positive correlation for Happyton, because the points seem to fall a little bit more closely near a straight line.

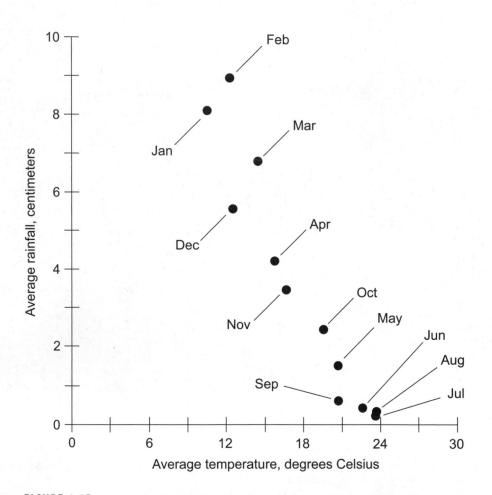

FIGURE 6-9B · Scatter plot of points showing the average monthly temperature versus the average monthly rainfall for Blissville.

Regression Curves

Curve fitting, which we learned about in Chap. 1, can illustrate the relationships among points in scatter plots such as those in Fig. 6-9A and B. Examples, based on "intuitive guessing," appear in Fig. 6-10. Figure 6-10A shows the same 12 points as those in Fig. 6-9A, representing the average monthly temperature and rainfall amounts for Happyton (without the labels for the months, to avoid cluttering things up). The dashed curve represents an approximation of a smooth function relating the two variables. In Fig. 6-10B, we do a similar curve-fitting exercise to approximate a function relating the average monthly temperature

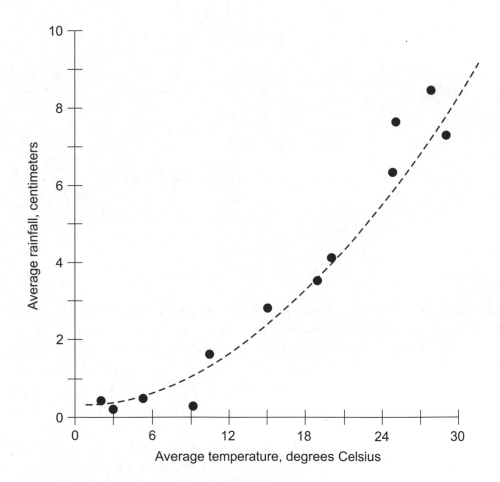

FIGURE 6-10A · Regression curve relating the average monthly temperature to the average monthly rainfall for Happyton.

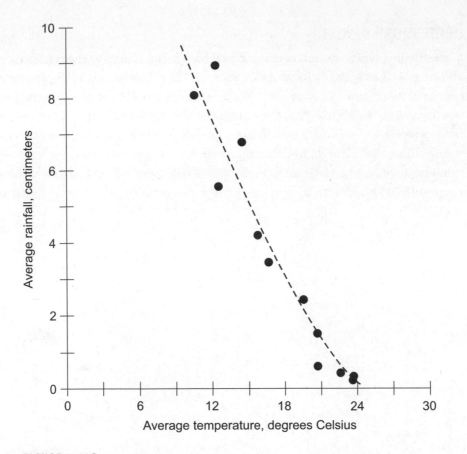

FIGURE 6-10B · Regression curve relating the average monthly temperature to the average monthly rainfall for Blissville.

and rainfall for Blissville. When we carry out a process of this sort, we determine the *regression*, and the curve that we get constitutes a *regression curve*.

In our hypothetical scenarios, we base the data shown in Table 6-1 and Figs. 6-8 through 6-10 on records gathered over 100 years. Now imagine that we had access to records gathered over a much longer time frame, say the past 1000 years! Further imagine that, instead of having data averaged by the month, we had data averaged by the week. In these cases, we would have to contend with gigantic tables, and we'd find the bar graphs impossible to read because of the clutter. But the scatter plots would tell a more interesting story. Instead of 12 points, each graph would have 52 points, one for each week of the year. We can reasonably suppose that the points would fall more closely along smooth curves than they do in Figs. 6-9 or 6-10.

Least-Squares Lines

As you might guess, computers can find ideal curves for scatter plots such as those shown in Fig. 6-9A and B. Most high-end scientific graphics suites include curve-fitting programs. We can also find computer programs that determine the best overall *straight-line* fit for the points in any scatter plot where correlation exists. Finding the ideal straight line is easier than finding the ideal smooth curve, although the result usually comes out less precise with the line than with the curve.

Figure 6-11A and B portrays the outputs of hypothetical computer programs designed to find the best straight-line approximations of the data from

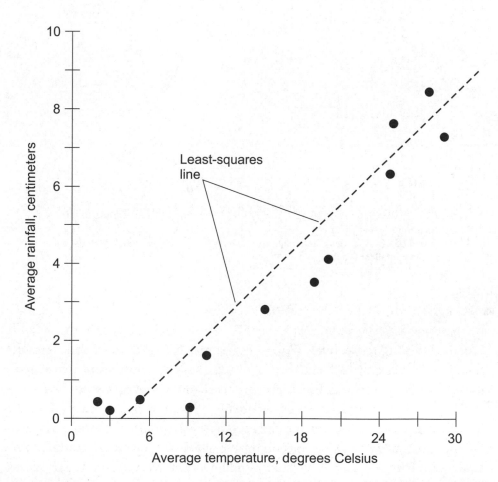

FIGURE 6-11A · Least-squares line relating the average monthly temperature to the average monthly rainfall for Happyton.

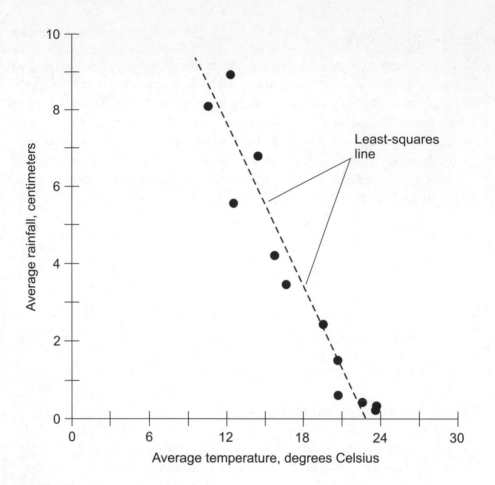

FIGURE 6-11B · Least-squares line relating the average monthly temperature to the average monthly rainfall for Blissville.

Fig. 6-9A and B, respectively. Suppose that the dashed lines in these graphs represent the best overall straight-line averages of the positions of the points. In that case, then the both lines obey a rule called the *law of least squares*.

We can find a *least-squares line* as follows. Suppose that we measure the distance between the dashed line and each of the 12 points in Fig. 6-11A. This process yields a set of 12 distance numbers, which we can call d_1 through d_{12}, all expressed in the same units (such as millimeters). We square these distance numbers, getting d_1^2 through d_{12}^2. Then we add these squared numbers up to obtain a final sum D. We'll find one *and only one* straight line for the scatter plot in Fig. 6-11A (or for any scatter plot where a correlation exists among the points) that produces a minimum value of D. That line constitutes the least-squares line. We can carry out the same process for the points and the dashed line in Fig. 6-11B.

When we have a scatter plot in which no correlation exists, we can't find a single least-squares line for the set of points. The calculation process produces ambiguous results. In this sort of situation, we could imagine that infinitely many least-squares lines exist. However, by convention in such a case, we say that no least-squares line exists.

Still Struggling

At this time you must wonder how anyone could muster the patience to work out least-squares lines for scatter plots involving large numbers of points. The answer: Computers, of course! A computer program designed to find the least-squares line for a scatter plot executes the aforementioned calculations and performs an *optimization problem*, calculating the equation of, and displaying, the line that best portrays the overall relationship among the points. Unless the points are haphazardly scattered or arranged in some coincidental fashion (such as uniformly placed around the perimeter of a circle or square) indicating that the correlation equals 0, such a program can quickly find and display the least-squares line for a scatter plot.

PROBLEM 6-5

Suppose the points in a scatter plot all lie exactly along a straight line so that the correlation equals either +1 (as strong as possible positively) or −1 (as strong as possible negatively). Where's the least-squares line in this type of scenario?

SOLUTION

If all the points in a scatter plot fall along a single straight line, then that line constitutes the least-squares line.

PROBLEM 6-6

Imagine that the points in a scatter plot lie all over the graph, so that the correlation equals 0. Where's the least-squares line in this case?

SOLUTION

When no correlation exists between two variables and the scatter plot shows this fact by the appearance of "randomly placed" points, then no least-squares line exists.

PROBLEM 6-7

Imagine that we've obtained the temperature versus rainfall data for the hypothetical towns of Happyton and Blissville (discussed earlier) on a daily basis rather than on a monthly basis. Also suppose that, instead of having been gathered over the past 100 years, the data was gathered over the past 1,000,000 years. We should expect, based on the more extensive data, to obtain scatter plots with points that lie almost perfectly along smooth lines or curves. We might want to use such data to express the present-day climates of the two towns. Why should we resist that temptation?

SOLUTION

The "million-year data" literally contains too much information to be useful in the present time. The earth's overall climate, as well as the climate in any particular location, has gone through wild changes over the past 1,000,000 years. Any climatologist, astronomer, or earth scientist can tell you that! Our planet has endured ice ages and warm interglacial periods; there have been wet and dry periods. While the 1,000,000-year data might be legitimate as it stands, it does not necessarily represent conditions this year, or last year, or over the past 100 years.

TIP   *Statisticians must avoid attempting to analyze too much information in a single experiment. Otherwise, the results can turn out skewed, or can produce "the right answer to the wrong question." When some crafty individual intends to introduce bias while making it look as if she has done an exceptionally good job of data collection, she might gather data over a needlessly large region or an unnecessarily long period of time. Beware!*

QUIZ

Refer to the text in this chapter if necessary. A good score is 8 correct. Answers are in the back of the book.

1. We can get a good idea of the nature and strength of the correlation between two variables by means of
   A. error correction.
   B. inference.
   C. null hypothesis.
   D. a scatter plot.

2. When we formulate an alternative hypothesis, we assume that
   A. no correlation exists between the variables in question.
   B. a specific null hypothesis is wrong.
   C. the data we have gathered is biased in some way.
   D. All of the above

3. Figure 6-12 shows a scatter plot between the values of two variables X and Y as determined by doing an experiment. We can tell by looking at this plot that the correlation between the variables is near or equal to
   A. 0.
   B. +1/2.
   C. +1.
   D. −1.

4. Suppose that we want to see how the average daily maximum temperature (in degrees Fahrenheit) correlates with the length of the daylight period (in minutes) at a certain location on the earth. We have the data for every single day of the calendar year, averaged over a period of 100 contiguous years (ignoring February 29 in leap years). The most practical and effective way to pictorially illustrate this correlation is to generate a
   A. median plot.
   B. scatter plot.
   C. bar graph.
   D. pie graph.

5. Figure 6-13 shows a scatter plot between the values of two variables W and Z as determined by doing an experiment. Suppose that we repeat the experiment and get identical results with one exception: The point at the extreme upper left, corresponding to approximately $W = 1.5$ and $Z = 8.4$ in the graph for the results of the first experiment, corresponds to approximately $W = 1.5$ and $Z = 6.1$ for the results of the second experiment. How does the least-squares line change in the graph for the results of the second experiment, compared with the graph for the results of the first experiment?

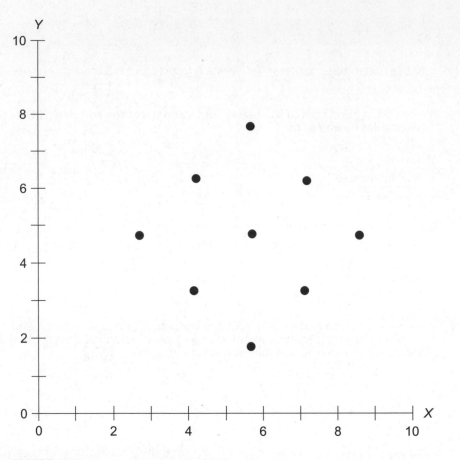

FIGURE 6-12 · Illustration for Quiz Question 3.

    A. The least-squares line ramps downward (as we move toward the right) a little less steeply in the scatter plot for the second experiment.

    B. The least-squares line ramps downward (as we move toward the right) a little more steeply in the scatter plot for the second experiment.

    C. The least-squares line does not change at all in the scatter plot for the second experiment, as compared with the scatter plot for the first experiment.

    D. The least-squares line acquires a small amount of curvature in the scatter plot for the second experiment.

6. If we consider W as the independent variable and Z as the dependent variable in Fig. 6-13, then the geometric slope (the "rise over run" as we move toward the right on the coordinate grid) of the least-squares line is

    A. negative.

    B. zero.

    C. positive.

    D. undefined.

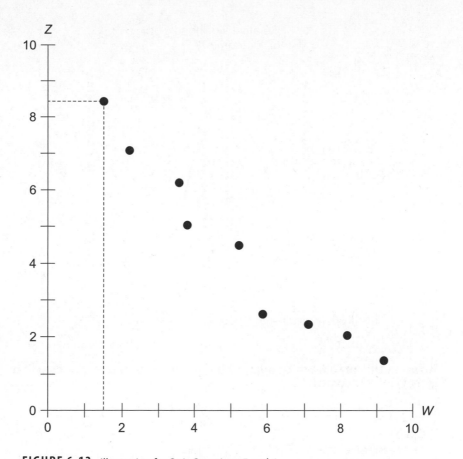

FIGURE 6-13 • Illustration for Quiz Questions 5 and 6.

7. Suppose that we use a computer program to "draw" a least-squares line for a scatter plot. Upon visual examination of the results, we can see that the points lie almost exactly along the least-squares line, which ramps steeply downward as we move toward the right in the graph. Evidently the correlation between the two variables is

A. larger than +1.
B. close to +1.
C. close to −1.
D. smaller than −1.

8. Figure 6-14 illustrates the relative probability that we'll observe a particular "outcome X" (vertical axis) as a function of the value of a specific independent variable (horizontal axis). We conduct five experiments over five different intervals representing defined spans of independent-variable values (vertical bars). Although the actual independent-variable values within each interval clearly differ, the intervals all have the same width, meaning that they all represent equally wide spans of values. Based on the information

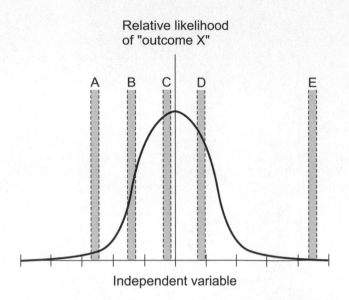

FIGURE 6-14 · Illustration for Quiz Questions 8 and 9.

shown in this graph, which span of independent-variable values will most likely yield "outcome X"?

A. Span A

B. Span B

C. Span C

D. We need more data to answer this question.

9. Based on the information shown in Fig. 6-14, which span of independent-variable values approximately indicates a 5% chance that we'll observe "outcome X"?

A. Span A

B. Span D

C. Span E

D. We need more data to answer this question.

10. The general process that we use to draw a conclusion on the basis of experimental data is known as

A. inference.

B. curve fitting.

C. correlation.

D. regression.

chapter 7

Correlation, Causation, Order, and Chaos

In this chapter, we'll see how cause/effect relationships can exist along with correlation. We'll also look at a couple of scenarios in which statistics borders on the theories of chaos, boundaries, and randomness.

CHAPTER OBJECTIVES

In this chapter, you will

- Determine the extent and nature of the correlation between variables
- Distinguish between qualitative and quantitative correlation
- Compare correlation, cause/effect, and coincidence
- Learn the basics of chaos theory and the butterfly effect
- Explore the nature of upper and lower bounds
- Attempt to define randomness

Correlation Principles

When we observe a correlation between two phenomena or effects, does one of them cause the other? Does a third phenomenon cause both? Does any cause/effect relationship exist? People often conclude that cause/effect (or *causation*) always associates with correlation. Sometimes it does. Sometimes it doesn't.

Quantitative versus Qualitative

We can *quantitatively* define correlation only between variables that we can express in terms of numerical values. Examples of quantitative variables include time, temperature, and average monthly rainfall. We can *qualitatively* express the correlation between two variables even if we can't express them as numerical values, but that's less precise, of course. Let's symbolize correlation as an italicized, lowercase letter r when we can quantify it.

Even if it seems obvious that two variables correlate, there's a big difference between saying, for example, "Bad manners and violence are strongly correlated" as opposed to claiming that "The correlation between rudeness and violence equals +0.75." We can quantify violence on the basis of numerical crime statistics, but we'll have trouble quantifying bad manners.

Imagine an experiment showing that people living in mountains develop "emotional problems" (such as anxiety, depression, and the like) somewhat less often than people living on prairies. We find both variables—emotional dysfunction and environment—difficult to quantify. We'll have trouble claiming that emotional problems and geographical elevation exhibit a correlation of, say, −0.47 or −0.38 or −0.29, even if the experiment demonstrates the existence of a negative correlation.

Correlation Range

As we've already learned, we can express quantitative correlation as a numerical value r in the range

$$-1 \leq r \leq +1$$

This inequality tells us that the mathematical correlation can equal anything between, and including, −1 and +1. We can use percentages if we like, so the possible range of correlation values $r_\%$ becomes

$$-100\% \leq r_\% \leq +100\%$$

A correlation value of $r = -1$ represents the strongest possible negative correlation, while $r = +1$ represents the strongest possible positive correlation. Moderately strong positive correlation will show up as something like $r = +0.7$; weak negative correlation will appear as something like $r_\% = -20\%$. A value of $r = 0$ or $r_\% = 0\%$ indicates the complete absence of correlation.

TIP *If we gather a sufficiently large number of samples (and place enough points on a scatter plot) when two phenomena or effects correlate, a computer will always come up with a definite positive or negative correlation figure. If no correlation exists, however, we'll find it difficult to prove that fact by quantitative "brute force," especially if we have a limited number of samples (or points in a scatter plot) to work with. A computer might yield a result of $r = +0.045$ or $r_\% = -0.87\%$, but rarely will a machine produce a mathematically zero correlation figure.*

?

Still Struggling

Two variables or effects can never correlate to an extent beyond the above-defined limits. If someone claims that two phenomena correlate by "a factor of -2," or if she says that "$r = 150\%$," you know that she's wrong. In addition, you must exercise caution when you say that two effects correlate to "twice the extent" of two other effects. If you know that two phenomena correlate by a factor of $r = +0.75$, and then someone comes along and tells you that changing the temperature (or some other parameter) will "double the correlation," you know that something is amiss; in such a case you'd have $r = +1.50$, an impossible scenario.

We Must Use a Straight Line

Quantitative correlation expresses the extent to which the points in a scatter plot fall near the least-squares line. We must pay attention to that key word "line"! If all the points lie along a straight line, then $r = +1$ if the line ramps upward to the right, or $r = -1$ if the line ramps downward to the right. From the principles you learned in first-year algebra and geometry, you can surmise the following two generalizations:

- If the values of both variables increase together, then $r > 0$, and the least-squares line has a *positive slope*, also called a positive "rise over run."

- If one value decreases as the other value increases, then $r < 0$, and the least-squares line has a *negative slope*, also called a negative "rise over run."

Once in awhile, you'll see a scatter plot in which all the points lie along a smooth curve that's not a straight line. In a case of that sort, one variable constitutes a mathematical function of the other. But points on a nonlinear curve can never exhibit a correlation of either −1 or +1. Figure 7-1A shows a scatter plot in which $r = +1$. Figure 7-1B shows a scatter plot with "perfect correlation" in the sense that the points lie along a smooth, definable curve, but $0 < r < +1$.

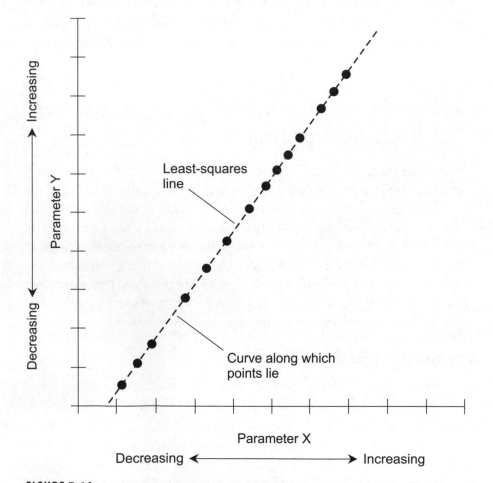

FIGURE 7-1A · Scatter plot in which the correlation equals +1. All the points lie along a straight line.

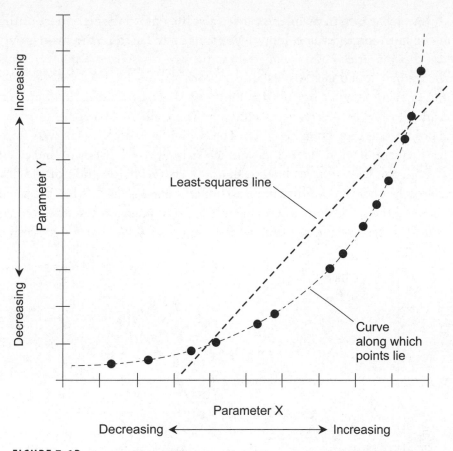

FIGURE 7-1B · Scatter plot in which the correlation does not equal +1. The points lie along a smooth curve but not a straight line.

TIP *Negative correlation translates to a least-squares line with negative slope, and positive correlation translates to a least-squares line with positive slope. However, the correlation is not the same parameter as the slope of the least-squares line! The correlation figure tells us how nearly the data points lie to the least-squares line representing them—whatever the slope of the line.*

Correlation and Outliers

In some scatter plots, the points appear near smooth curves or lines, although we'll rarely encounter a scatter plot with points as orderly as those shown in

Fig. 7-1A or B. Once in awhile, we'll see a scatter plot in which most of the points lie near a straight line, but a few points stray far from the main group. We call the "maverick points" *outliers*. In some ways, these "stray" points resemble the outliers found in statistical distributions.

One or two outliers can (but don't always) affect the correlation between two variables. Figure 7-2 shows a scatter plot where all but two points lie in the same positions as they do in Fig. 7-1A. The two outliers both lie far away from the least-squares line! In Fig. 7-2, the outliers exist at equal distances (indicated by d) from the least-squares line, so their effects on the position of the line "cancel each other out" so that the least-squares line in Fig. 7-2 has the same position and the same slope as the one in Fig. 7-1A. Nevertheless, the correlation values differ. We can see straightaway from Fig. 7-1A that $r = +1$. In the

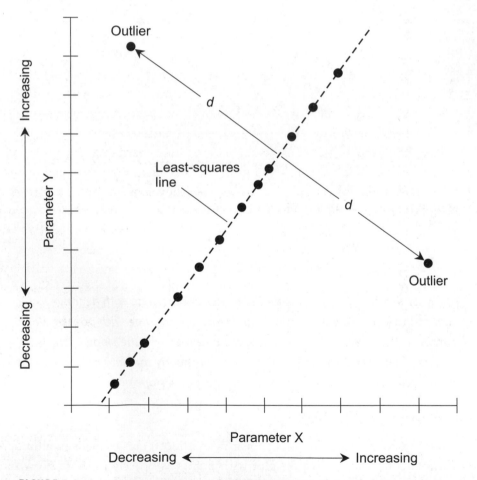

FIGURE 7-2 · Scatter plot in which two outliers reduce the correlation, even though they don't affect the position of the least-squares line.

situation shown by Fig. 7-2, we know that $r < +1$ because not all of the points fall along the least-squares line.

Correlation and Definition of Variables

For the purposes of determining correlation, it doesn't matter which variable we define as dependent and which variable we define as independent. If we transpose only the definitions of the variables but nothing about the "real-world" scenario changes, the correlation, once we calculate it, comes out exactly the same.

Think back to the previous chapter, where we analyzed the correlation between average monthly temperatures and average monthly rainfall amounts for two cities. When we generated the scatter plots, we plotted the temperatures on the horizontal axes and called the temperatures the independent variables. However, we could have plotted the rainfall amounts on the horizontal axes and defined them as the independent variables instead. The resulting scatter plots would have looked a lot different, but upon mathematical analysis, the correlation figures wouldn't have been any different.

Sometimes a particular quantity or phenomenon, such as time or geographic location, lends itself intuitively to the role of the independent variable. In the cases of Happyton and Blissville from the previous chapter, it doesn't matter much which variable we call independent and which variable we call dependent. In fact, these labels can mislead because they suggest the existence of a cause/effect relation.

Do temperature variations, over the course of the year, influence the rainfall in Happyton or Blissville? If so, the effects differ—in fact, they act in the opposite sense—between the two cities. How about the reverse cause/effect relation, where rainfall amounts influence the temperature? Again, if that's true, the effect in one city operates "in reverse" compared to the effect in the other town. Both of these two cause/effect hypotheses lead us to rather weird conclusions. Perhaps another factor, or even a combination of multiple factors, influences both the temperature and the rainfall.

Units (Usually) Don't Matter

Here's an interesting property of correlation. It doesn't make any difference how large or small our measurement units might be, as long as they express the applicable phenomena or characteristics for the variables. If the measurement unit of either variable changes in size but not in essence, the appearance of a bar graph or scatter plot changes. The plot "stretches out" or "squashes in" vertically or horizontally. But the correlation figure, r, between the two variables remains unaffected.

Think back again to the last chapter, and the scatter plots of precipitation versus temperature for Happyton and Blissville. We render the precipitation amounts in centimeters per month, and we express the temperatures in degrees Celsius. Suppose that we render the precipitation amounts in inches per month instead? The graphs will look a little different, but upon analysis by a computer, the correlation figures will turn out the same. Suppose that we express the temperatures in degrees Fahrenheit? Again, the graphs will look different, but r won't change. Even if we plot the average rainfall in miles per month and the temperatures in Kelvins (where 0 K represents absolute zero, the coldest possible temperature), the value of r will come out the same.

TIP *We can change* unit size *when we apply the foregoing rule, but we must never alter the quantities or phenomena that the units represent. If we plot the average rainfall in inches, centimeters, or miles* per week*, we can't expect the correlation to work out the same as they do when we go in inches, centimeters, or miles* per month*. The scatter plots will no longer portray the same functions, because the variable on the vertical scale—rainfall averaged over weekly periods rather than over monthly periods—will no longer represent the same observed phenomenon.*

PROBLEM 7-1

Suppose that we cut the distances of both outliers from the least-squares line in Fig. 7-2 in half (to $d/2$ rather than d), as shown in Fig. 7-3. How will this action affect the correlation? How will it affect the location and slope of the least-squares line?

SOLUTION

The correlation will increase, because the average distance between all the points and the least-squares line will decrease. However, the least-squares line won't move.

PROBLEM 7-2

Suppose that we remove one of the outliers in the scenario of Fig. 7-3. Will this action affect the location and slope of the least-squares line? If so, how?

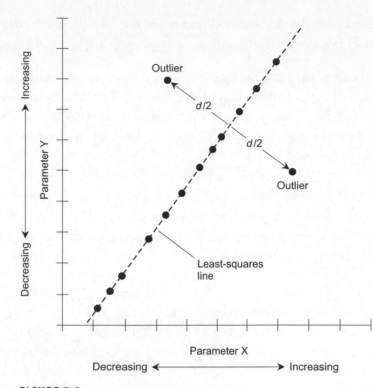

FIGURE 7-3 • Illustration for Problems 7-1 and 7-2.

✔️ **SOLUTION**

If we take away the upper-left outlier, the least-squares line will move slightly downward and to the right, and its slope will decrease a tiny bit. If we take away the lower-right outlier, the least-squares line will move slightly upward and to the left, and its slope will increase a tiny bit.

Cause and Effect

Whenever we find that a significant correlation exists between two variables, we tend to believe that a cause/effect relationship must exist as well. Examples abound; nobody who listens to the radio, reads newspapers, browses the Internet, or watches television can escape them. Sometimes people suggest or insinuate cause/effect relationships without directly claiming them. For example:

- Take one of these pills every morning, and you'll smile all the time.
- Avoid sweet foods, and you'll never die of a heart attack.

- Drink this bitter liquid, and you'll never get dirt under your toenails.
- Eat this gray goo, and the hair on your ears will vanish within 2 weeks.

Scams of this sort always give themselves away; we can always rewrite them in the general form "Do something that causes me to make money, and then your life will improve." Sometimes a cause/effect relationship actually exists in situations like this. Sometimes it doesn't. Often, we don't know—and can't easily find out, either!

Correlation and Causation

Let's reduce a correlation situation to generic terms. That way, we won't inappropriately infer causation. Suppose that two phenomena, called X and Y, vary in intensity with time. Figure 7-4 shows a relative graph of the

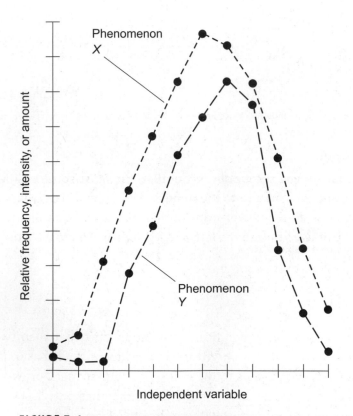

FIGURE 7-4 • The two phenomena shown here, X and Y, appear to correlate strongly. Does the correlation imply cause/effect?

fluctuations in both phenomena. They change in a manner that's positively correlated. When X increases, so does Y, in general. When Y decreases, so does X, in general.

Cause/effect can occur in the situation of Fig. 7-4 in several different ways, but we must always consider the hypothesis that no causation exists. Maybe Fig. 7-4 shows a sheer coincidence. If we had 1000 points on each plot, we'd have a better case for causation. But we have only 12 points on each plot. Do these points represent a "freak scenario"? In the "real world," we must also consider a more sinister hypothesis whenever we see plots that contain only a few points, such as these have. We must ask ourselves, "Did someone with a vested interest in the outcome of the analysis carefully select the 12 points in each plot of Fig. 7-4, out of many hundreds or even thousands of available points, to create the appearance of a correlation between X and Y when in fact no correlation exists whatsoever?

When we assign real phenomena or observations to the variables in a graph such as Fig. 7-4, we can get ideas about causation. But these ideas don't necessarily reflect reality; they can even mislead us. Intense debate often takes place in scientific, political, and religious circles concerning whether or not a correlation between two variables reflects cause/effect—and if so, how the causative relation or function actually operates. In the examples that follow, let's rule out the bias factor and assume that all data has been obtained with the intent of pursuing truth.

X Causes Y

We can illustrate cause/effect relationships using arrows. Figure 7-5A shows the situation where changes in phenomenon X directly cause changes in phenomenon Y. You can doubtless think of many scenarios.

Suppose that the independent variable, shown on the horizontal axis in Fig. 7-4, represents the time of day between sunrise and sunset. Plot X shows the relative intensity of sunshine during this time period; plot Y shows the relative temperature over that same period of time. We can argue that the brilliance of the sunshine causes the changes in temperature. We see some time lag in the temperature function; this observation should not surprise us. The hottest part of a clear day usually comes a little while after "high noon" when the sunlight has the greatest intensity.

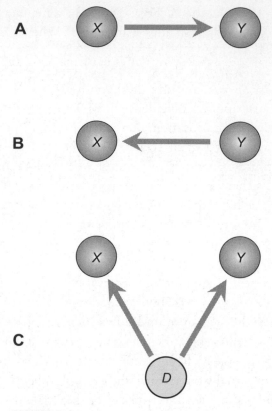

FIGURE 7-5 • At A, *X* causes *Y*. At B, *Y* causes *X*. At C, *D* causes both *X* and *Y*.

? Still Struggling

We can—with difficulty—imagine a cause/effect relationship in the other direction. Suppose that heating causes the sky to completely clear of clouds, resulting in the maximum possible amount of sunlight at the surface (*Y* causes *X*). Such an argument makes a "twisted" sort of sense, but most meteorologists would say that the former relation better represents reality.

Y Causes X

Imagine that the horizontal axis in Fig. 7-4 represents 12 different groups of people in a medical research survey. Each hash mark on the horizontal axis represents one group. Plot X shows a point-to-point graph of the relative number of fatal strokes in a given year for the people in each of the 12 groups; plot Y shows a point-to-point graph of the relative average blood pressure levels of the people in the 12 groups during the same year. (These are hypothetical graphs, not based on actual experiments, but a "real-life" survey might come up with results like this. Medical research has shown a correlation between blood pressure and the frequency of fatal strokes.)

Do we have a cause/effect relationship between the value of X and the value of Y here? Most doctors would answer with a qualified yes: variations in Y cause the observed variations in X (Fig. 7-5B). Stated simply, high blood pressure can cause fatal strokes in the sense that, if all other factors remain constant, a person with high blood pressure will more likely suffer a fatal stroke than another person with normal blood pressure.

X Causes Y

What about the reverse cause/effect hypothesis in the above-described scenario? Can fatal strokes cause high blood pressure (X causes Y)? If so, time would have to "flow" in reverse, wouldn't it?

Complications

Of course, the foregoing situations are oversimplifications. The cause/effect relationships described aren't "pure." In real life, "pure" cause/effect events, where we see a single defined cause and a single inevitable effect, rarely occur.

The brightness of the sunshine is not, all by itself, the only cause/effect factor in the temperature during the course of a day. A nearby lake or ocean, the wind direction and speed, and the passage of a weather front can all influence the temperature at any given location. We've all seen the weather clear and brighten up, along with an abrupt drop in temperature, when a strong front passes by. The sun comes out, and at the same time a chilly wind rushes in. An event like that contradicts the notion that bright sun causes things to heat up, even though the simplistic theoretical argument remains valid "when all other factors remain constant." Problems arise when one or more "other factors" change!

In regards to the blood-pressure-versus-stroke relationship, numerous other variables come into play, and scientists don't necessarily know them all. New discoveries constantly emerge in the medical field. Examples of other factors that might play cause/effect roles in the occurrence of fatal strokes include age, heredity, nutrition, emotional well-being, blood cholesterol level, and body fat index. A cause/effect relationship (Y causes X) exists, but it's not "pure."

Something Else Causes Both X and Y

Now suppose that the horizontal axis in Fig. 7-4 represents 12 different groups of people in another medical research survey. Again, each hash mark on the horizontal axis represents one group. Plot X is a point-to-point graph of the relative number of heart attacks in a given year for the people in each of the 12 groups; plot Y is a point-to-point graph of the relative average blood cholesterol levels of the people in the 12 groups during the same year. As in the stroke scenario, these are hypothetical but plausible graphs. Medical science has demonstrated the existence of a correlation between blood cholesterol and the frequency of heart attacks.

When doctors and their students first began examining the bodies of people who died of heart attacks in the middle 1900s, they found "lumps" called *plaques* in the arteries. Scientists theorized that plaques cause blood clots that can cut off the circulation to parts of the heart, causing tissue death. The plaques contain cholesterol. Evidently, cholesterol can accumulate inside the arteries. When the scientists saw data showing a correlation between blood cholesterol levels and heart attacks, they hypothesized that if the level of cholesterol in the blood could be reduced, the likelihood of the person having a heart attack later in life would go down.

Heart specialists began telling their patients to eat fewer cholesterol-containing foods, hoping that this dietary change would reduce blood cholesterol levels. In many cases, that indeed happened. Evidence accumulated to support the hypothesis that a low-cholesterol diet reduces the likelihood that a person will have a heart attack. More than mere correlation operates here. There's causation, too. Between what variables, and in what directions, does the causation operate?

Let's call the amount of dietary cholesterol "factor D." According to current medical theory, there exists a cause/effect relation between factors D and X, and there also exists a correlation between factors D and Y. Some studies have indicated that, all other things being equal, people who eat lots of cholesterol-rich food items (or who consume a lot of cholesterol) have more heart attacks than people who eat few such food items (or consume little cholesterol). Figure 7-5C illustrates the scenario. Most nutritionists and physicians will tell you that a cause/effect relation exists between factor D (the amount of cholesterol in the diet) and factor X (the number of heart attacks); a cause/effect relation also exists between factor D and factor Y (the average blood cholesterol level). But most scientists would also agree that this portrayal fails to represent the entire situation in the "real world." If you adopt a strict vegetarian diet and avoid cholesterol-containing foods altogether, you do not thereby derive any guarantee that you'll never experience a heart attack. If you eat steak and eggs every day for breakfast, it doesn't automatically follow that you're doomed to have a heart attack. The cause/effect relationship exists, but it's not "pure," and it's not absolute.

Multiple Factors Cause Both X and Y

If you watch television shows with advertisements aimed at middle-aged and older folks, you'll hear all about cholesterol and heart disease—maybe more than you want to hear. High cholesterol, low cholesterol, HDL, LDL, big particles, small particles. You might start wondering whether you should go to a chemistry lab rather than a kitchen to prepare your food. The cause/effect relationship between cholesterol and heart disease is complicated. The more we learn, it appears, the less we know.

Let's introduce and identify three new variables. Factor S represents "stress" (in the sense of anxiety and frustration), factor H represents "heredity" (in the sense of genetic background), and factor E represents "exercise" (in the sense of physical activity). Over the past several decades, scientists have suggested the existence of cause/effect relationships between each of these factors and blood cholesterol levels, and between each of these factors and the frequency of heart attacks. Figure 7-6 illustrates this sort of "cause/effect web." Proving the validity of each link—for example, whether or not stress, all by itself, can influence cholesterol in the blood—remains a task for future researchers.

Coincidence

The existence of correlation between two phenomena doesn't necessarily imply any particular cause/effect scenario. Two phenomena can correlate because of a sheer coincidence. Two phenomena can appear to correlate even if in fact they do not, especially when we attempt to derive conclusions from insufficient or inappropriate data. In the case of blood cholesterol versus the frequency of heart attacks, test populations have traditionally contained thousands of elements (people). The researchers can therefore rest assured in their conclusions

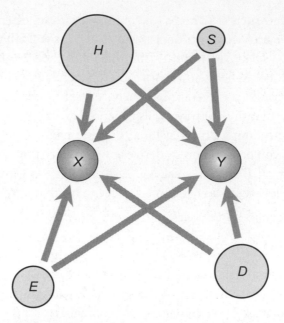

FIGURE 7-6 • A scenario with multiple causes *D, S, H*, and *E* for both *X* and *Y*, not necessarily all having equal effect or importance.

that such correlation does not represent the product of coincidence. We can reasonably suppose that some cause/effect interaction exists. Researchers are still busy figuring out exactly how it all works, and if they ever succeed in completely unraveling the mystery, it's a good bet that an illustration of the "cause/effect web" will look more complicated than Fig. 7-6.

Correlation indicates that things take place more or less in concert with one another, allowing us to predict certain future events with varying degrees of accuracy. But another, much different form of order exists in nature. This peculiar behavior, defined by a new science called *chaos theory*, illustrates that some phenomena, totally unpredictable and which defy statistical analysis in the short term or on a small scale, can nevertheless exhibit a high degree of order and predictability on a large scale.

PROBLEM **7-3**

What are some reasonable cause/effect relationships that might exist in Fig. 7-6, other than those shown or those directly between *X* and *Y*? Use arrows to show cause/effect, and use the abbreviations shown.

✔ **SOLUTION**

Consider the following hypotheses. Think about how you might conduct statistical experiments to check the validity of these notions, and how you might determine the extent of the correlation.

- H → S (Hypothesis: Some people are born more stress-prone than others.)
- H → D (Hypothesis: People of different genetic backgrounds have developed cultures where the diets dramatically differ.)
- E → S (Hypothesis: Regular, frequent exercise can relieve or reduce stress.)
- E → D (Hypothesis: Extreme physical activity makes people eat more food, particularly carbohydrates, because those people need more food.)
- D → S (Hypothesis: Bad nutritional habits can worsen stress. Consider a hard-working person who lives on coffee and potato chips, versus a hard-working person who eats mainly fish, fruits, vegetables, and whole grains.)
- D → E (Hypothesis: People with good nutritional habits get more exercise than people with bad nutritional habits.)

PROBLEM 7-4

What are some cause/effect relationships in the diagram of Fig. 7-6 that are questionable or absurd?

✔ **SOLUTION**

Consider the following hypotheses. Think about how you might conduct statistical experiments to find out whether or not you can move the first three items into the preceding category.

- H → E (Question: Do people of certain genetic backgrounds naturally get more physical exercise than people of other genetic backgrounds?)
- S → E (Question: Can stress motivate some types of people to exercise more, yet motivate others to exercise less?)
- S → D (Question: Do certain types of people eat more under stress, while others eat less?)
- S → H (Isn't this idea ridiculous? Stress can't affect a person's heredity!)
- E → H (Isn't this idea ridiculous? Exercise can't affect a person's heredity!)
- D → H (Isn't this idea ridiculous? Dietary habits can't affect a person's heredity!)

Chaos, Bounds, and Randomness

Have you ever noticed that events—especially dramatic or catastrophic events—tend to occur in bunches? A few decades ago, the engineer and mathematician *Benoit Mandelbrot* observed and analyzed this effect. His work gave birth to the science of *chaos theory*.

Event Bunching

In the early summer of 1992, south Florida hadn't experienced a severe hurricane since Betsy in 1965. The area around Miami gets hit by a "full-blown" hurricane once every 7 or 8 years on the average, and an extreme storm once or twice a century. Was Miami "due" for a hurricane in the early '90s? Had the time arrived for a massive event? Some people said yes. The year 1992 was no more or less special, in terms of hurricane probability, than any other year. In fact, as the hurricane season began in June of that year, experts predicted a season of below-normal activity, and that prediction proved accurate in general—with one big glitch. On August 24, 1992, Hurricane Andrew tore across the southern suburbs of Miami and the Everglades, proving itself the costliest hurricane ever to hit the United States up to that date.

Did Andrew's unusual intensity have anything to do with the lack of hurricanes during the previous two and a half decades? No. Did Andrew's passage make a similar event in 1993 or 1994 less likely than it would have been if Andrew had not hit south Florida in 1992? No. There could have been another storm like Andrew in 1993, and two more in 1994. Theoretically, a half dozen more storms like Andrew could have taken place later in 1992!

Have you ever heard about a tornado hitting some town, followed 3 days later by another one in the same region, and 4 days later by another, and a week later by still another? Have you ever flipped a coin for a few minutes and had it come up "heads" a dozen times in a row, even though you'd normally have to flip it for days to expect such a thing to happen? Have you witnessed some vivid example of event bunching, and wondered if anyone has ever come up with a mathematical theorem that explains why this sort of thing seems so common?

Slumps and Spurts

Athletes such as competitive swimmers and runners know that improvement characteristically comes in spurts, not smoothly with the passage of time.

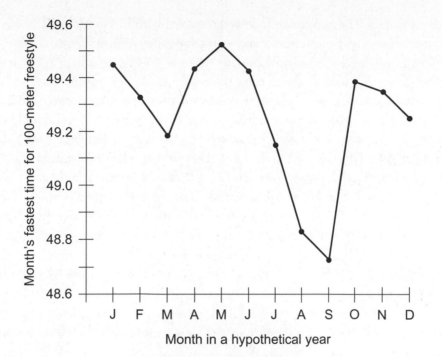

FIGURE 7-7 · Monthly best times (in seconds) for a swimmer in the 100-meter freestyle race, plotted by month for a hypothetical year. Illustration for Problems 7-5 and 7-6.

Figure 7-7 shows an example as a graph of the date (by months during a hypothetical year) versus time (in seconds) for a hypothetical athlete's 100-meter freestyle swim. The horizontal scale shows the month, and the vertical scale shows the swimmer's fastest time in that month.

Note that the swimmer's performance does not improve for awhile, and then suddenly it does. In this example, almost all of the improvement occurs during June, July, and August. Another swimmer might exhibit performance that worsens during the same training season. Do such irregularities imply that all of the training done during times of flat performance represents time wasted? The coach will say no! Why does improvement take place in sudden bursts, and not gradually with time? Sports experts will tell you they don't know. We see similar effects in the growth of plants and children, in the performance of corporate sales departments, and in the frequency with which people get sick.

Correlation, Coincidence, and Chaos

Sometime during the middle of the 20th century, a researcher noticed a strong correlation between the sales of television sets and the incidence of heart attacks in Great Britain. The two curves followed remarkably similar contours. The shapes of the graphs were, peak-for-peak and valley-for-valley, almost identical. Why?

Here's one theory: As people bought more television sets, they spent more time sitting and staring at the screens; the resulting idleness meant that they got less exercise; the people's physical condition therefore deteriorated, making them more susceptible to heart attacks. But this argument, if valid, couldn't explain the *uncanny exactness* with which the two curves followed each other, year after year. There would have been a lag effect if television watching really did cause poor health, but no lag occurred. It was as if a person could go out and buy a television set and reasonably expect to suffer the ill effects *straightaway*.

Do television sets emit energy fields that cause immediate susceptibility to a heart attack? Is the programming so terrible that it causes immediate physical harm to viewers? Both of these notions seem "far-out." Were the curves obtained by the British researcher nearly identical for some unsuspected reason? Or was the whole thing a coincidence that eluded all attempts at explanation? Did no true correlation exist at all between television sales and heart attacks, a fact that would have eventually emerged had the experiment continued for decades longer or involved more people?

? Still Struggling

Scientists sometimes search for nonexistent cause/effect explanations, getting more puzzled and frustrated as the statistical data pours in, demonstrating the existence of a correlation but giving no clue as to what is responsible for it. Applied to economic and social theory, this sort of correlation-without-causation phenomenon can lead to some frightening propositions. Are we all doomed to fight another world war, endure another economic disaster, or suffer through another disease pandemic because that's "simply how things are"?

Scale-Recurrent Patterns

Benoit Mandelbrot noticed that patterns tend to recur or duplicate over widely diverse time scales. Large-scale and long-range changes take place in patterns similar to those of small-scale and short-term changes. Events occur in bunches; the bunches themselves take place in similar bunches following similar patterns. This effect exists both in the increasing scale and in the decreasing scale.

Have you noticed that high, cirrostratus clouds in the sky resemble the clouds in a room where someone has recently lit up a cigar, or that these clouds look eerily like the interstellar gas-and-dust clouds that make up diffuse nebulae in space? Patterns in the physical universe often fit inside each other like nested geometric shapes, as if the repetition of patterns over scale takes place because of some principle ingrained in nature itself. You can find such patterns when you look at the so-called *Mandelbrot set* (Fig. 7-8) using any of numerous zooming programs available on the Internet. This set arises from a *simple, finite* mathematical formula but exhibits *infinite complexity*. No matter how much

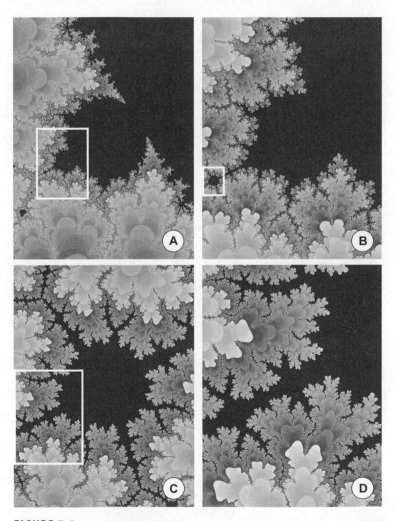

FIGURE 7-8 · Portions of the Mandelbrot set as seen with increasing levels of "magnification" (views enlarge progressively from A through D).

you magnify the Mandelbrot set—that is, however closely you zoom in on it—new patterns appear. Nevertheless, the patterns show similarity at all scales.

The images in Fig. 7-8 were generated with a freeware program called *Fractint*. At the time of writing, this program was created by a group of programming experts called the Stone Soup Team. The program itself bears a copyright, but images created by any user become the property of that user.

PROBLEM 7-5

Does the tendency of athletic performance to occur in "spurts" mean that a gradual improvement can never take place in any real-life situation? For example, could the curve for the swimmer's times (Fig. 7-7) look like either the solid or dashed lines in Fig. 7-9 instead?

SOLUTION

Gradual, smooth improvement in athletic performance can take place. Either of the graphs (the straight, dashed line or the solid, smooth curve)

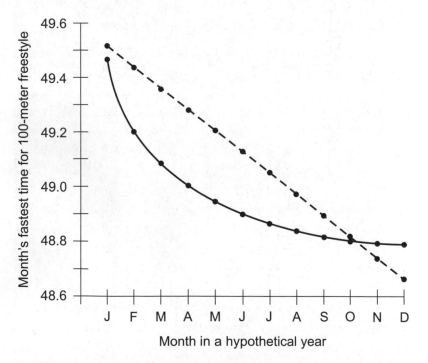

FIGURE 7-9 · Illustration for Problems 7-5 and 7-6.

in Fig. 7-9 could represent a real-life situation. But orderly improvement scenarios such as the ones graphed in Fig. 7-9 occur less often than the more chaotic type of variation such as Fig. 7-7 portrays.

PROBLEM 7-6

What's the fastest possible time for the 100-meter freestyle swimming event, as implied by the graphs in Fig. 7-7 or Fig. 7-9?

SOLUTION

Neither of these graphs logically implies the existence of any specific time representing the fastest possible 100-meter swim. Mathematicians can, however, prove that there exists a *maximum unswimmable time* for this event. Determining the actual time as a specific number, and then proving that the determination holds valid, constitutes another problem altogether.

The Maximum Unswimmable Time

If our hypothetical swimmer keeps training, how fast will he eventually swim the 100-meter freestyle? We already know that he can do it in a little more than 48 seconds. What about 47 seconds? Or 46 seconds? Or 45 seconds?

We can find obvious *lower bounds* to the time in which a human can swim the 100-meter freestyle. Any reasonable coach would agree that nobody will ever carry off this event in 10 seconds. How about 11 seconds, then? Or 12 seconds? Or 13 seconds? Still ridiculous? How about 20 seconds? Or 25 seconds? Or 30 seconds? If we start at some impossible figure such as 10 seconds and keep increasing the number gradually, we will at some point reach a figure—let's suppose for the sake of argument that it's 41—representing the largest whole number of seconds too fast for anyone to swim the 100-meter freestyle.

Once we have two whole numbers, one representing a swimmable time (say 42 seconds) and the next smaller one representing an unswimmable time (say 41 seconds), we can refine the process down to the tenth of a second, and then to the hundredth, and so on indefinitely. There exists some value, exact to however small a fraction of a second we care to express it, representing the maximum unswimmable time (MUST) that a human being can attain for the 100-meter freestyle swim. Figure 7-10 shows an educated guess for this situation.

No one knows the exact MUST for the 100-meter freestyle swimming event, and we might argue that no human being (or computer) can precisely determine it. But such a time nevertheless exists. How do we know that a specific MUST, in seconds, exists for the 100-meter freestyle, or for any other event in any other timed sport? A well-known theorem of mathematics, called the *theorem of the greatest lower bound*, makes it plain: "If there exists a lower bound for a set, then there exists a *greatest lower bound* for that set." A more technical term for "greatest lower bound" is *infimum*. In this case, the set in question is the set of "swimmable times" for the 100-meter freestyle. The lower bounds are the "unswimmable times."

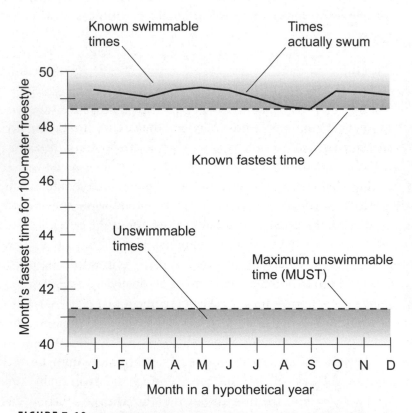

FIGURE 7-10 • According to the theorem of the greatest lower bound, there exists a maximum unswimmable time (MUST) for the 100-meter freestyle. (The time shown here constitutes a mere guess.)

? Still Struggling

What's the probability that a human being will come to within a given number of seconds of the MUST for the 100-meter freestyle in, say, the next 10 years, or 20 years, or 50 years? Sports writers will speculate on it; sports doctors will come up with ideas; swimmers and coaches doubtless have notions too. But anyone who makes a claim in this respect makes a guess, and nothing but a guess. We can't say "The probability equals 50% that someone will swim the 100-meter freestyle in so-and-so seconds by the year such-and-such." For any theoretically attainable time, say 43.50 seconds, one *and only one* of two things will happen: Either someone will swim the 100-meter freestyle that fast someday, or else no one will do it.

The Butterfly Effect

The tendency for small events to have dramatic long-term and large-scale consequences has a catchy name: the *butterfly effect*. This expression arises from a hypothetical question: Can a butterfly taking off in China affect the development, intensity, and course of a hurricane 6 months later in Florida? At first, such a question seems outlandish. But imagine that the butterfly creates a tiny air disturbance that produces a slightly larger one, and so on, and so on, and so on. According to butterfly-effect believers, the insect's momentary behavior could constitute the trigger that ultimately dictates the difference between a rain shower and a killer cyclone thousands of kilometers away and many days later.

We can never know all the consequences of any particular event. History happens once, *and only once*. We can't make repeated trips back and forth through time and let fate unravel itself over and over after "tweaking" this, or that, or the other little detail. But events can conspire, or have causative effects over time and space, in such a manner as to magnify the significance of tiny events in some circumstances. Scientists have programmed computers to demonstrate the phenomenon.

Suppose that you go out cycling in the rain and subsequently catch a cold. The cold develops into pneumonia, and you barely survive. Might things have turned out differently if the temperature had been a little warmer, or if it had rained a little less, or if you had stayed out for a little less time? No practical algorithm exists that can determine which of these tiny factors are critical and which are not. But we can set up computer models, and we can run programs, that in effect "replay history" with various parameters adjusted. In some cases, certain variables exhibit threshold points where a tiny change right now will dramatically affect the distant future.

Scale Parallels

In models of chaos, patterns repeat in large and small sizes for an astonishing variety of phenomena. Compare, for example, the image of a spiral galaxy as viewed through a large telescope with the image of a hurricane as seen from an earth-orbiting satellite. The galaxy's stars compare to the hurricane's water droplets. The galaxy's spiral arms compare to the hurricane's rainbands. In a hurricane, everything rushes in toward the eye as if it were a black hole. The water droplets, carried by winds, spiral inward more and more rapidly as they approach the edge of the eye. In a spiral galaxy, the stars move faster and faster as they fall inward toward the center, which may indeed contain a cosmic black hole.

Both pressure and gravitation can, as they operate over time and space on a large scale, produce the same general form of spiral. You'll find similar spirals in the Mandelbrot set and other mathematically derived patterns. For example, the *Spiral of Archimedes* (a standard spiral easily definable in analytic geometry) occurs often in nature, and in widely differing scenarios.

? Still Struggling

We can convince ourselves that the above-described structural parallels represent something more than sheer coincidence—that a cause/effect relationship must exist. But what cause/effect factor can make a spiral galaxy in outer space appear and behave so much like a hurricane on the surface of the earth? Does gravitation in the galaxy function in the same sense as the air pressure in the hurricane? Ask your science teacher!

PROBLEM 7-7

Can the MUST scenario, in which a greatest lower bound exists, apply in a reverse sense? For example, might a *minimum unattainable temperature (MUAT)* exist for our planet?

✓ SOLUTION

The highest recorded temperature on earth, as of this writing, is approximately 58°C (136°F). Given current climatic conditions, we can easily "invent" an unattainable temperature, for example, 500°C. We can then start working our way down from this figure. Clearly, 400°C is unattainable, as is 300°C, and also 200°C (assuming runaway global warming doesn't take place, in which our planet might end up with an atmosphere like that of Venus). What about 80°C? What about 75°C? A theorem of mathematics, called the *theorem of the least upper bound*, makes the situation plain: "If there exists an upper bound for a set, then there exists a *least upper bound* for that set." It follows that some MUAT for our planet must exist, given current climatic conditions.

The Malthusian Model

Some people apply chaos theory to portray doomsday characteristics of the earth's population growth. Suppose that we want to find a function to describe world population versus time. The simplest model allows for an exponential increase in population, but this so-called *Malthusian model* (named after its alleged author *Thomas Malthus*, an English priest and scientist who lived in the 18th and 19th centuries) fails to incorporate factors such as disease pandemics, world wars, or the collision of an asteroid with our planet.

The Malthusian model begins with the notion that as time passes, the world's human population will increase exponentially, while the world's available supply of food and other resources will increase arithmetically. It follows that a pure Malthusian population increase can only go on for a certain length of time, and then crippling shortages will force a slowdown. When the population reaches a certain critical maximum, the population will no longer increase, because the earth will become too crowded and will lack sufficient resources to keep everyone alive for a normal lifespan.

What will happen then? Will the population level off smoothly? Will it decline suddenly and then increase again? Will it decline gradually and then stay low? The outcome depends on the values we assign to certain parameters in the function that describes population versus time.

A Bumpy Ride

The limiting process for any population-versus-time function depends on the extent of the disparity between population growth and resource growth. If we consider the earth's resources finite, then the shape of the population-versus-time curve depends on how fast people reproduce until a catastrophe occurs. As the reproduction rate goes up—as we "drive the function harder"—the time period until the first crisis decreases, and the ensuing fluctuations become wilder. Malthusian population growth takes place according to the formula

$$x_{n+1} = rx_n (1 - x_n)$$

where n represents an increasing whole number starting with $n = 0$, and r represents a factor that quantifies the rate of population increase. (Don't confuse this with the r factor that represents correlation, defined earlier in this chapter!)

Statisticians, social scientists, biologists, mathematicians, and even some politicians have run the Malthusian population formula through computers for various values of r to get an idea what would happen to the world's population as a function of time on the basis of various degrees of "population growth pressure." As things work out, a leveling-off condition occurs when the value of r remains less than about 2.5. As the value of the r factor increases, it "drives the function harder," and the population increases with greater rapidity until a certain point in time—and then chaos erupts.

According to computer models, when the r factor remains low, the world population increases, reaches a peak, and then falls back. Then the population increases again, reaches another peak, and undergoes another decline. This takes place over and over, but with gradually diminishing wildness. The catastrophes, however they might manifest themselves, become less and less severe. Mathematicians would say that a *damped oscillation* occurs in the population function as it settles down to a *steady state* (Fig. 7-11A).

In real life, humanity might keep the r factor low by means of strict population control through public education. Conversely, the r factor could increase if all efforts at population control fail. Computers tell us with unwavering display screens what they "think" will happen in that case. If r rises to a sufficient value, the ultimate world population will not settle down. Instead, it will oscillate indefinitely between limiting values as shown in Fig. 7-11B. The amplitude and frequency of the oscillation will depend on how large the r factor becomes.

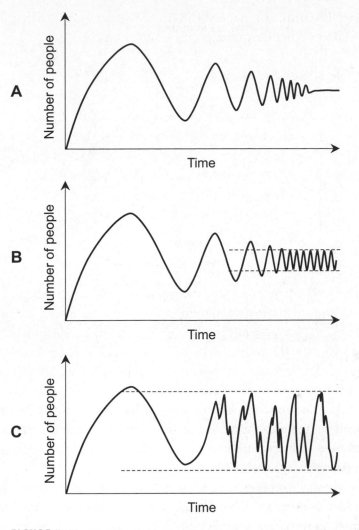

FIGURE 7-11 • World human population versus time. At A, a small *r* factor produces eventual stabilization. At B, a large *r* factor produces oscillation within limits. At C, a huge *r* factor produces chaotic fluctuation within limits.

At a large enough critical value for the *r* factor, the population-versus-time function fluctuates crazily, never settling down to any apparent oscillation frequency, although the population "peaks and valleys" might remain within definable bounds (Fig. 7-11C).

A graph in which we plot the world's ultimate human population on the vertical (dependent-variable) axis and the *r* factor on the horizontal (independent-variable) axis produces a characteristic pattern something like the one shown

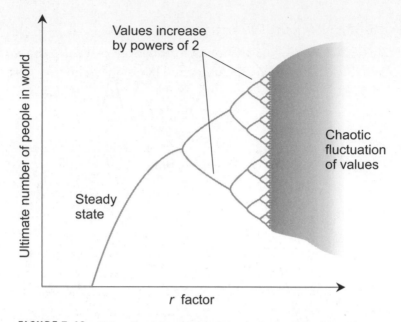

FIGURE 7-12 • Generalized, hypothetical graph showing "final" world population as a function of the relative r factor.

in Fig. 7-12. The function breaks into oscillation when the r factor reaches a certain value. At first this oscillation has defined frequency and amplitude. But as r continues to increase, we reach a point at and beyond which the oscillation turns into "noise" (chaos).

Think about what happens when you ramp up the audio gain (volume control) of a public-address or music-band sound system until feedback from the speakers finds its way to the microphone, and the speakers begin to howl. If you increase the audio gain some more, the oscillations get louder. If you drive the system harder still, the oscillations increase in fury until, when you crank the gain up all the way to the top, the system roars like thunder. That situation would also produce a graph like the one in Fig. 7-12, where the horizontal axis shows the setting of the gain control and the vertical axis shows the level of the sound at the speakers.

Does the final population figure in the right-hand part of Fig. 7-12 truly represent unpredictable variation between extremes? If the computer models represent reality, then chaos experts will say yes. By all indications, the gray area in the right-hand part of Fig. 7-12 portrays something like a state of "randomness." We can only hope that our world never enters a "gray area" like that.

The "Randomness" Conundrum

Statisticians occasionally want to obtain sequences of values that occur at random. What constitutes true randomness? Here's one definition that we can apply to single-digit numbers:

- A sequence of digits from the set {0, 1, 2, 3, 4, 5, 6, 7, 8, 9} is *truly random* if and only if, given any digit in the sequence, there exists no way for a human or machine to predict the next one.

At first thought, this definition makes it seem that the task of generating a sequence of truly random numbers should be easy. Suppose that we chatter away out loud, aimlessly uttering the names of single-digit numbers from 0 to 9, and record our voice on a computer? That scheme won't work, because any individual person has a "leaning" or preference for certain digits or sequences of digits, such as 5 or 58 or 289 or 8827. I have such a preference, and so do you. We may not know what it is, but if we were willing to submit to an intensive and lengthy enough study, we could find out.

If a sequence of digits proceeds in truly random fashion, then over long periods a given digit x will occur exactly 10% of the time, a given sequence xy will occur exactly 1% of the time, a given sequence xyz will occur exactly 0.1% of the time, and a given sequence $wxyz$ will occur exactly 0.01% of the time. These percentages should hold for all possible sequences of digits of the given sizes, and similar rules should hold for sequences of any length. But if you speak or write down digits for a few days and record the result, you can have almost complete confidence that things won't work out that way.

Here's another definition of true randomness. We base this definition on the hypothesis that all artificial processes contain inherent orderliness:

- A sequence of digits is *truly random* if and only if there exists no algorithm capable of generating the next digit in a sequence, on the basis of the digits already generated in that sequence.

According to this definition, if we can show that any digit in a sequence constitutes a function of those before it, then we can have complete confidence that the sequence *is not* truly random. We can rule out many sequences that seem random to the casual observer. For example, we can generate the value of the square root of 2 (or $2^{1/2}$) with an algorithm called *extraction of the square root*. We can apply this algorithm to any whole number that's not a perfect square. If we have the patience, and if we know the first n digits of a square root, we can find the $(n + 1)$st digit by means of this process. It works every time, and we get the same result every time. The digits in the decimal expansion of the square root of 2, as well as the decimal expansion of any other irrational number, emerge in the same sequence—*exactly* the same sequence—every time a computer grinds it out. The decimal expansions of irrational numbers therefore do not give us truly random digit sequences according to the above definition.

If the digits in any given irrational number fail to occur in a truly random sequence, where can we find digits that do? Can such a sequence even exist within the realm of "real-world thought"? If we can't generate a random sequence of digits using an algorithm, does this fact rule out any thought process that allows us to identify the digits? Are we looking for something so elusive that, when we think we've finally discovered it, the very fact that we've gone through a thought process to find it proves that we have not? If that's true, then how can statisticians get hold of a random sequence that they can employ to their satisfaction?

In the interest of practicality, statisticians sometimes settle for *pseudorandom* digits or numbers. The prefix *pseudo-* in this context means "for all intents and purposes." We can find plenty of algorithms that generate strings of pseudorandom digits.

The Web to the Rescue

You can search the Internet and find sites with information about pseudorandom and random numbers. You'll find a few downloadable programs that can turn a home computer into a generator of pseudorandom digits. Go to your favorite search engine, bring up the page that allows you to conduct an advanced or word-sequence search, and then enter the phrase "random number generator." Exercise caution when you download any executable program. Make sure that your antivirus software is effective and up to date. If you're uneasy about downloading programs from the Web, then don't do it. If your Web browser issues any sort of warning when you prepare to download a file, heed that warning.

You can get random digits from a Web site maintained by the well-known and respected mathematician Dr. Mads Haahr of Trinity College in Dublin, Ireland. Point your browser to *www.random.org*. The author describes the difference between pseudorandom and truly random numbers. He also links to sites for further research. Dr. Haahr's Web site makes use of *electromagnetic noise* (something like the "static" that thunderstorms produce on the radio) to obtain real-time random-number sequences. For this scheme to work, there must exist no sources of orderly noise near enough to be picked up by the receiver. Orderly noise sources include internal combustion engines and certain types of electrical appliances such as old light dimmers.

PROBLEM 7-8

Name two poor sources of random digits, and two good sources, that we can obtain by practical means.

SOLUTION

The digits of a known irrational number such as the square root of 20 represent an example of a poor source of random digits. These digits are *predestined* (they already exist, and they come out the same every time we generate them for any particular irrational number).

The repeated spinning of a wheel or rotatable pointer, calibrated in digits from 0 to 9 around the circumference, gives us another example of a poor way to get a random sequence. We can never spin the wheel forcefully enough to "randomize" the final result. In addition, the amount of mechanical friction will likely vary at different points in the wheel bearing's rotation, causing it to favor certain digits over others.

We can obtain a good source of random digits by building a special die that has the shape of a *regular dodecahedron*, a geometric solid with 12 identical faces. We can number the faces from 0 to 9, leaving two faces blank. We can repeatedly toss this "hyper-die" and note the digit, if any, that appears on the straight-upward-facing side. (We ignore tosses that produce a blank face.) We can tally the results over time as random digits. The die must have uniform density throughout, to ensure that it doesn't favor some digits over others.

Another good source of random digits is a machine similar to the type used in televised lottery drawings. Powerful air jets blow a set of 10 light-weight balls, numbered 0 through 9, around inside a large pressurized bottle. To choose a digit, the "lottery judge" opens a small lid on top of the bottle just long enough to let one of the balls fly out. After the "judge" notes the digit on the "snagged" ball, she returns the ball to the jar, and the 10 balls fly around for a little while before the "judge" repeats the process to get another digit.

QUIZ

Refer to the text in this chapter if necessary. A good score is 8 correct. Answers are in the back of the book.

1. According to the Malthusian hypothesis, the supply of resources available to humanity, as time passes, will
    A. eventually run out.
    B. increase up to a certain maximum.
    C. increase arithmetically.
    D. decrease logarithmically.

2. With strong negative correlation, the points in a scatter plot lie
    A. near a straight line.
    B. near the dependent-variable axis.
    C. near the independent-variable axis.
    D. all over the coordinate grid.

3. Refer to the scatter plot of Fig. 7-13. Suppose that the dashed line represents the least-squares line for all the solid black points. If we add a new value at point *P* but don't change anything else, what will happen to the *slope* of the least-squares line?
    A. It will not change.
    B. It will remain negative and get steeper.
    C. It will remain negative and get less steep.
    D. It will become positive.

4. Refer to the scatter plot of Fig. 7-13. Suppose that the dashed line represents the least-squares line for all the solid black points. If we add a new value at point *Q* but don't change anything else, what will happen to the *slope* of the least-squares line?
    A. It will not change.
    B. It will remain negative and get steeper.
    C. It will remain negative and get less steep.
    D. It will become positive.

5. Refer to the scatter plot of Fig. 7-13. Suppose that the dashed line represents the least-squares line for all the solid black points. If we add a new value at point *R* but don't change anything else, what will happen to the *slope* of the least-squares line?
    A. It will not change.
    B. It will remain negative and get steeper.
    C. It will remain negative and get less steep.
    D. It will become positive.

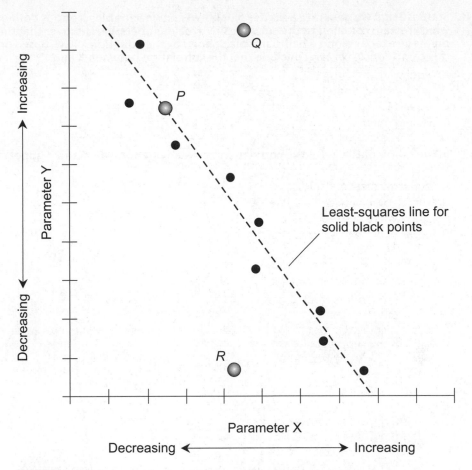

FIGURE 7-13 • Illustration for Quiz Questions 3 through 5.

6. **If we find that the correlation between two phenomena *X* and *Y* equals +50%, then *in general*, an increase in *X* will**

   A. have an unexpected effect on *Y*.
   B. have no effect on *Y*.
   C. go along with a decrease in *Y*.
   D. go along with an increase in *Y*.

7. **When a seemingly insignificant event produces a dramatic change in the out-come of an experiment, we witness an example of**

   A. strong correlation.
   B. chaotic correlation.
   C. butterfly effect.
   D. true randomness.

8. Suppose that we generate a scatter plot between two variables X and Y, both of which remain confined to the set of positive real numbers. We discover that the points lie exactly along a half-parabola defined by the equation $Y = X^2$. Based on this coincidence, we can conclude that the correlation between X and Y

   A. equals +1.
   B. is between 0 and +1.
   C. equals 0.
   D. is undefined.

9. In the graph of Fig. 7-14, the correlation between phenomena X and Y appears to be

   A. equal or close to +100%.
   B. somewhere between 0 and +100%.
   C. somewhere between 0 and −100%.
   D. equal or close to −100%.

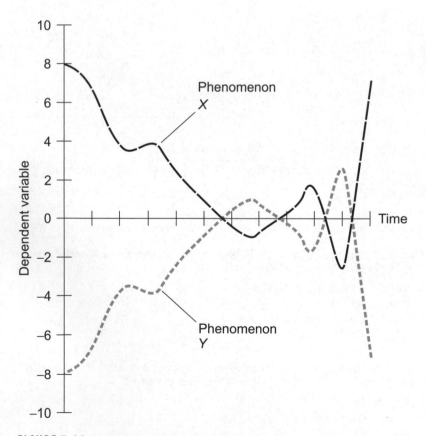

FIGURE 7-14 • Illustration for Quiz Question 9.

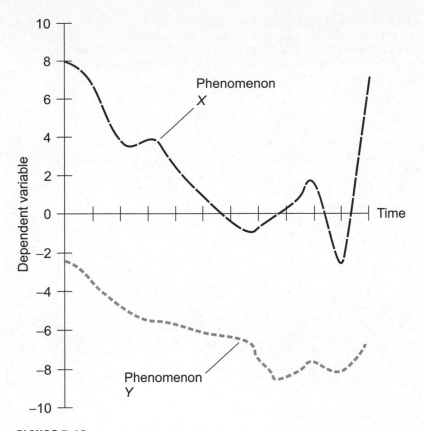

FIGURE 7-15 • Illustration for Quiz Question 10.

10. **In the graph of Fig. 7-15, the correlation between phenomena *X* and *Y* appears to be**
    A. equal or close to +100%.
    B. somewhere between 0 and +100%.
    C. somewhere between 0 and −100%.
    D. equal or close to −100%.

chapter 8

Practical Review Problems

Let's statistically analyze some hypothetical "real-world" scenarios. The scope of this chapter encompasses all the material in the preceding chapters.

CHAPTER OBJECTIVES: REVIEW YOUR KNOWLEDGE

In this chapter, you will review your knowledge about

- Frequency distributions
- Variance
- Standard deviation
- Probability
- Data intervals
- Sampling
- Estimation
- Hypotheses
- Prediction
- Regression
- Correlation
- Cause/effect relations

Frequency Distributions

Imagine that we give a quiz to a large class of students. We examine the results in the form of tables and graphs.

PROBLEM 8-1

In our class, 130 students take a 10-question quiz. We're told the following facts:

- Nobody missed all the questions
- Four people got one correct answer
- Seven people got two correct answers
- Ten people got three correct answers
- Fifteen people got four correct answers
- Twenty-four people got five correct answers
- Twenty-two people got six correct answers
- Twenty-four people got seven correct answers
- Fifteen people got eight correct answers
- Seven people got nine correct answers
- Two people wrote perfect papers

Portray these results in the form of a table, showing the test scores in ascending order from top to bottom in the left-hand column and the absolute frequencies for each score in the right-hand column.

SOLUTION

Table 8-1 shows the results of the quiz in tabular form. In this depiction, the lowest score appears at the top and the highest score appears at the bottom, according to the instructions we got.

PROBLEM 8-2

How else can we arrange the data from Problem 8-1 in tabular form?

SOLUTION

We can portray the quiz results in a table with the highest score at the top and the lowest score at the bottom (Table 8-2). We can also interchange the columns and rows, so that the table has two rows and 11 columns (not

TABLE 8-1 Table for Problem 8-1. The lowest score appears at the top and the highest score appears at the bottom.

Test Score	Absolute Frequency
0	0
1	4
2	7
3	10
4	15
5	24
6	22
7	24
8	15
9	7
10	2

TABLE 8-2 Table for Problem 8-2. Here, we see the same data as Table 8-1 portrays, but with the highest score at the top and the lowest score at the bottom.

Test Score	Absolute Frequency
10	2
9	7
8	15
7	24
6	22
5	24
4	15
3	10
2	7
1	4
0	0

counting the column with the headers). We can create this arrangement in either of two ways: the lowest score at the left and the highest score at the right (Table 8-3A), or the highest score at the left and the lowest score at the right (Table 8-3B).

TABLE 8-3A   Another table for Problem 8-2. We see the same data as Table 8-1 shows, but in a horizontal configuration rather than a vertical configuration. The lowest score appears at the left and the highest score appears at the right.

Test Score	0	1	2	3	4	5	6	7	8	9	10
Absolute Frequency	0	4	7	10	15	24	22	24	15	7	2

TABLE 8-3B   Still another table for Problem 8-2. We see the same data as Table 8-3A portrays, but with the highest score at the left and the lowest score at the right.

Test Score	10	9	8	7	6	5	4	3	2	1	0
Absolute Frequency	2	7	15	24	22	24	15	10	7	4	0

PROBLEM 8-3

Render the data from Problem 8-1 in the form of a vertical bar graph, showing the lowest score at the left and the highest score at the right. Don't put numbers for the absolute frequency values at the tops of the bars.

SOLUTION

Figure 8-1 portrays the results of the quiz as a vertical bar graph, without absolute frequency values shown at the tops of the bars. We alternate the shading (light and dark) to help us (and our viewers) distinguish clearly between adjacent bars.

PROBLEM 8-4

Render the data from Problem 8-1 in the form of a horizontal bar graph, showing the lowest score at the top and the highest score at the bottom. Include the absolute frequency values at the right-hand ends of the bars.

SOLUTION

Figure 8-2 shows the results of the quiz as a horizontal bar graph, with the absolute frequency values at the right-hand ends of the bars. We alternate the shading for clarity.

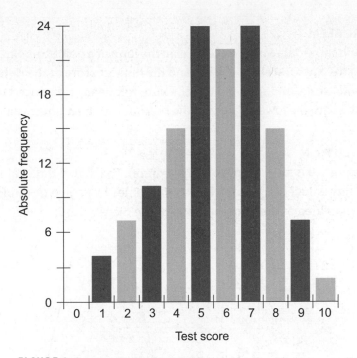

FIGURE 8-1 · Illustration for Problem 8-3.

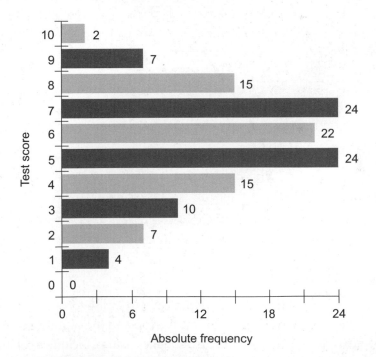

FIGURE 8-2 · Illustration for Problem 8-4.

PROBLEM **8-5**

Render the data from Problem 8-1 in the form of a point-to-point graph, showing the lowest score at the left and the highest score at the right on the horizontal scale, and showing the absolute frequency referenced to the vertical scale with the lowest values at the bottom and the highest values at the top.

SOLUTION

Figure 8-3 is an example of such a graph. The actual data values appear as points (small black dots). The straight lines help give us (and our viewers) a good idea of the general shape of the distribution.

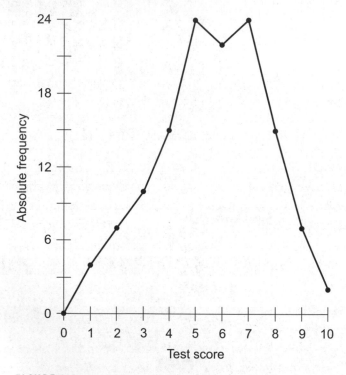

FIGURE **8-3** • Illustration for Problem 8-5.

PROBLEM **8-6**

Portray the results of our hypothetical quiz in the form of a table similar to Table 8-1, with the lowest score at the top and the highest score at the bottom. In addition to the absolute frequency values, include a column showing cumulative absolute frequencies in ascending order from top to bottom.

TABLE 8-4 Table for Problem 8-6. The cumulative absolute frequency values constantly increase as we read downward.

Test Score	Absolute Frequency	Cumulative Absolute Frequency
0	0	0
1	4	4
2	7	11
3	10	21
4	15	36
5	24	60
6	22	82
7	24	106
8	15	121
9	7	128
10	2	130

SOLUTION

See Table 8-4. The values in the third column (cumulative absolute frequency) always increase as we go down the table. That is, each number exceeds the one above it. In addition, the largest value equals the total number of elements in the statistical group, in this case 130 (the number of students in the class).

PROBLEM 8-7

Render the data from Problem 8-1 in the form of a dual point-to-point graph, showing the lowest score at the left and the highest score at the right on the horizontal scale. Show the absolute frequency values as a set of small black dots connected by solid lines, referenced to a vertical scale at the left-hand side of the graph. Show the cumulative absolute frequency values as a set of small open circles connected by dashed lines, referenced to a vertical scale at the right-hand side of the graph.

SOLUTION

See Fig. 8-4. The solid black dots and the solid black line go with the left-hand scale; the open circles and the dashed line go with the right-hand scale.

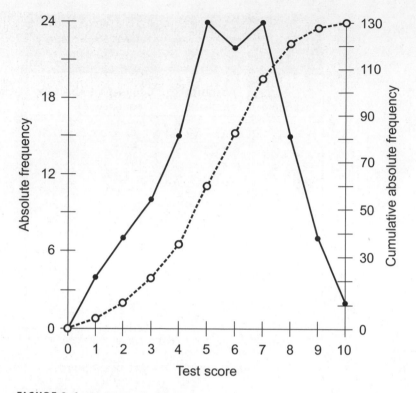

FIGURE 8-4 · Illustration for Problem 8-7.

PROBLEM 8-8

What's the population mean for the quiz results in our class?

SOLUTION

To find the population mean, we multiply each score by its absolute frequency, obtaining a set of products. Then we add up the products and divide the result by the number of papers in the class, in this case 130. Table 8-5 shows the products, along with the cumulative sums. Making a table and then double-checking the results can help us avoid errors in situations like this. (If a mistake occurs, it propagates through the rest of the calculation and gets multiplied, worsening the inaccuracy of the final result.) The population mean equals 731/130, or approximately 5.623. We symbolize it as μ.

TABLE 8-5 Table for Problem 8-8. The lowest score appears at the top and the highest score appears at the bottom. We divide number at the lower right by the number of elements in the population (in this case 130) to obtain the mean, which works out to approximately 5.623.

Test Score	Absolute Frequency	Abs. Freq. × Score	Cum. Sum of Products
0	0	0	0
1	4	4	4
2	7	14	18
3	10	30	48
4	15	60	108
5	24	120	228
6	22	132	360
7	24	168	528
8	15	120	648
9	7	63	711
10	2	20	731

PROBLEM 8-9

What's the median for the quiz results in our class?

SOLUTION

As we learned in Chap. 2, the median for the discrete variable in a distribution constitutes the value such that the number of elements greater than or equal to it matches the number of elements less than or equal to it. In this particular example, the "elements" are the test results for each individual in the class. Table 8-6 shows how we go about finding the median. When we tally up the scores of all 130 individual papers so that they appear in order, we find that the scores of the 65th and 66th papers—the two in the middle—are both six correct answers. We can conclude that the median equals 6, because half of the students got six or more answers correct, and half of the students got six or fewer answers correct.

TABLE 8-6 Table for Problem 8-9. We can determine the median by tabulating the cumulative absolute frequencies.

Test Score	Absolute Frequency	Cumulative Absolute Frequency
0	0	0
1	4	4
2	7	11
3	10	21
4	15	36
5	24	60
6 (partial)	5	65
6 (partial)	17	82
7	24	106
8	15	121
9	7	128
10	2	130

PROBLEM 8-10

What's the mode score for the quiz results in our class?

SOLUTION

We define the mode as the score that occurs most often, or the set of scores that occurs most often (if we end up with a redundant result). In this situation we have two mode scores: five correct answers and seven correct answers. We therefore have a bimodal distribution in which the values equal 5 and 7.

PROBLEM 8-11

Show the mean, median, and modes for our test as vertical dashed lines in a point-to-point graph similar to the plot of Fig. 8-3.

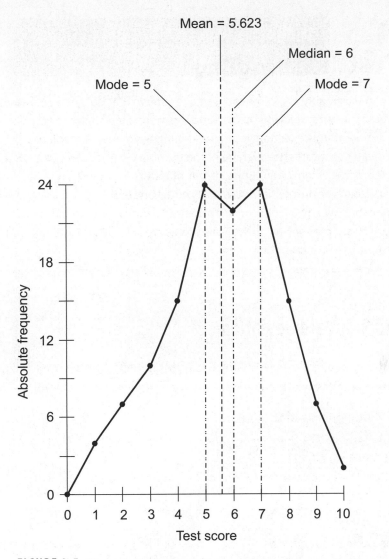

FIGURE 8-5 • Illustration for Problem 8-11.

SOLUTION _____

See Fig. 8-5. We've labeled the mean, median, and modes, referencing them all to the horizontal scale.

? Still Struggling

You can reasonably ask, "How can we have a mean that doesn't equal any of the scores in the quiz, while the median and the mode correspond to actual scores?" The answer to this question lies in the definitions of mean, median, and mode. The mean constitutes the average of the numbers in a set. Obviously that average won't necessarily equal one of the numbers in the set (in fact, it usually won't). However, the median will usually equal one of the scores; the only exception is when two adjacent scores "compete." The mode, assuming that we don't have a "freak scenario" in which all the scores are achieved by identical numbers of students, will always equal at least one of the scores.

Variance and Standard Deviation

Let's further analyze the results of the hypothetical 10-question quiz given to the class of 130 students. Again, in verbal form, here are the results:

- Nobody missed all the questions
- Four people got one correct answer
- Seven people got two correct answers
- Ten people got three correct answers
- Fifteen people got four correct answers
- Twenty-four people got five correct answers
- Twenty-two people got six correct answers
- Twenty-four people got seven correct answers
- Fifteen people got eight correct answers
- Seven people got nine correct answers
- Two people wrote perfect papers

PROBLEM 8-12

What's the variance of the distribution of scores for the hypothetical quiz whose results appear as detailed above? Round off the answer to two decimal places.

✔️ SOLUTION

Let x represent the independent variable. We have $n = 130$ individual quiz scores, with one score for each student. Let's call these individual scores by the names x_i, where i represents an *index number* that can range from 1 to 130 inclusive. Therefore, we have 130 x_i's ranging from x_1 to x_{130}. Let's call the absolute frequency for each particular numerical score f_j, where j represents an index number that can range from 0 to 10 inclusive, so we have 11 f_j's ranging from f_0 to f_{10}. The population mean, μ, equals approximately 5.623, as we've already found. We want to calculate the variance of x, symbolized Var(x). We learned in Chap. 2 that we can calculate the variance of a set of n values x_1 through x_n, with a population mean of μ, with the formula

$$\text{Var}(x) = (1/n)\,[(x_1 - \mu)^2 + (x_2 - \mu)^2 + \ldots + (x_n - \mu)^2]$$

Let's compile a table as we work through the above formula, and fill in the table values as we make calculations. In Table 8-7, we list all of the possible test scores in the first (far-left-hand) column. In the second column, we show

TABLE 8-7 Table for Problem 8-12. The population mean, μ_p, equals approximately 5.623. We symbolize individual test scores as x_i. Because the class contains 130 students, i ranges from 1 to 130 inclusive. We denote the absolute frequency for each score as f_j. We have 11 possible scores; we allow j to range from 0 through 10 inclusive.

Test Score x_i	Abs. Freq. f_j	$x_i - 5.623$	$(x_i - 5.623)^2$	$f_j(x_i - 5.623)^2$	Cum. Sum of $f_j(x_i - 5.623)^2$ Values
0	0	−5.623	31.62	0.00	0.00
1	4	−4.623	21.37	85.48	85.48
2	7	−3.623	13.13	91.91	177.39
3	10	−2.623	6.88	68.80	246.19
4	15	−1.623	2.63	39.45	285.64
5	24	−0.623	0.39	9.36	295.00
6	22	0.377	0.14	3.08	298.08
7	24	1.377	1.90	45.60	343.68
8	15	2.377	5.65	84.75	428.43
9	7	3.377	11.40	79.80	508.23
10	2	4.377	19.16	38.32	546.55

the absolute frequency for each score. In the third column, we subtract the population mean, 5.623, from the score, obtaining a tabulation of the differences between μ and each particular score x_i. In the fourth column, we square each of these differences. In the fifth column, we multiply the numbers in the fourth column by the absolute frequencies in the second column. In the sixth column, we compile cumulative sums the numbers in the fifth column, getting a grand total of 546.55 at the lower-right corner of the table. We multiply this quantity by $1/n$ (or divide it by n) to get the variance. In our case, $n = 130$. Therefore, rounding off to two decimal places, we have

$$\text{Var}(x) = 546.55/130$$
$$= 4.20$$

PROBLEM 8-13

What's the standard deviation of the distribution of scores for the quiz? Round off the answer to two decimal places.

SOLUTION

The standard deviation, symbolized σ, equals the square root of the variance. We already know that the variance equals 4.20, rounded off to two decimal places. We can find the square root using a calculator. To minimize the risk of encountering any rounding error in our calculations, let's take the square root of the original quotient to obtain

$$\sigma = (546.55/130)^{1/2}$$
$$= 2.05$$

PROBLEM 8-14

Draw a point-to-point graph of the distribution of quiz scores in the situation that we've dealt with in the past few problems and solutions, showing the mean and indicating the range of scores within one standard deviation of the mean ($\mu \pm \sigma$).

SOLUTION

First, let's calculate the values of the mean minus the standard deviation ($\mu - \sigma$) and the mean plus the standard deviation ($\mu + \sigma$). We get

$$\mu - \sigma = 5.623 - 2.050 = 3.573$$

and

$$\mu + \sigma = 5.623 + 2.050 = 7.673$$

Figure 8-6 shows these values, along with the mean itself. The scores that fall within this range appear as small circles with light interiors. The scores that fall outside this range appear as small black dots.

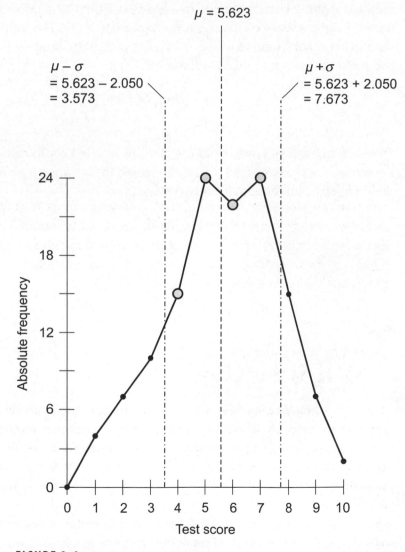

FIGURE 8-6 · Illustration for Problem 8-14.

PROBLEM 8-15

In the quiz scenario we've analyzed, how many students got scores within one standard deviation of the mean? What percentage of the total students does this span represent? Round the answer off to the nearest percentage point.

SOLUTION

The scores that fall within the range $\mu \pm \sigma$ are 4, 5, 6, and 7. Referring to the original data portrayed in Table 8-1, we can see that 15 students got four answers right, 24 students got five answers right, 22 students got six answers right, and 24 students got seven answers right. The total number (let's call it n_σ) of students who got answers within the range $\mu \pm \sigma$ equals the sum of these numbers, as follows:

$$n_\sigma = 15 + 24 + 22 + 24$$
$$= 85$$

The total number of students in the class, n, equals 130. To calculate the percentage of students (call it $n_{\sigma\%}$) who scored in the range $\mu \pm \sigma$, we must divide n_σ by n and then multiply by 100, getting

$$n_{\sigma\%} = 100 \, (n_\sigma/n)$$
$$= 100 \, (85/130)$$
$$= 65\%$$

Still Struggling

When we solve problems such as those in the last few paragraphs, we can easily lose our grasp of what it all means! Students of statistics sometimes complain about "loss of bearing." We have formulas, numbers, and calculators. We input the numbers into the calculators and punch buttons according to the formulas, getting output values. Then we forget what our answer means! In a case like that, we should review the applicable definition and try to think of it in plain language. For example, in the solution we just completed, we calculated the percentage of students that scored within a certain range above or below the average. In slang, we can say that we defined the "middle of the pack."

Probability

The next several problems deal with situations and games involving chance. Probability lends itself to "thought experiments" that don't require any material investment except a few sheets of paper, a pen, and a calculator.

PROBLEM 8-16

Consider a "fair" die with 12 faces, in the shape of a regular dodecahedron. For any single toss, the probability of the die coming up on any face equals the probability of its coming up on any other face. (For our purposes, the term "coming up" means that a face is parallel to the table onto which we throw the die and in plain sight, not lying against the table.) The faces bear the numerals 1 through 12. What's the probability that the die will come up showing the numeral 5 twice in a row? How about three times in a row, or four times in a row, or n times in a row (where n represents any whole number)?

SOLUTION

The probability P_1 that the dodecahedral die will come up on any particular face (including the one with the numeral 5), on any particular toss, equals 1/12. The probability P_2 that the die will come up showing the numeral 5 twice in a row equals $(1/12)^2$, or 1/144. The probability P_3 that the die will come up showing the numeral 5 three times in a row equals $(1/12)^3$, or 1/1728. The probability P_4 that the die will come up showing the numeral 5 four times in a row equals $(1/12)^4$, or 1/20,736. In general, the probability P_n that a 12-faced die will come up showing the numeral 5 exactly n times in a row equals $(1/12)^n$. Each time we toss the die, the chance of adding to our consecutive string of 5's equals exactly 1 in 12.

PROBLEM 8-17

Imagine a set of "fair" dodecahedral dice. Suppose that we throw a pair of them simultaneously. What's the probability that both dice will come up showing 3? Now suppose that we throw three of these dice at the same time. What's the probability that all three of them will come up showing 3? How about four simultaneously tossed dodecahedral dice all coming up 3? How about n simultaneously tossed octahedral dice all coming up 3?

✔ SOLUTION

The progression of probabilities here is the same as that for a single die tossed over and over. The probability P_1 that the die will come up on any particular face (including 3), on a single toss, equals 1/12. The probability P_2 that a pair of dodecahedral dice will both come up on any particular face, such as 3, equals $(1/12)^2$, or 1/144. The probability P_3 that three dice will come up showing 3 equals $(1/12)^3$, or 1/1728. The probability P_4 that four dice will come up showing 3 equals $(1/12)^4$, or 1/20,736. In general, the probability P_n that n dodecahedral dice, tossed simultaneously, will all come up showing 3 equals $(1/12)^n$.

◻ PROBLEM 8-18

Imagine that an election takes place for a seat in the state senate. In our hypothetical district, most of the residents lean toward "politically liberal." Two candidates take part in this election, one of whom calls herself "liberal" and who has been in office for many years, and the other of whom calls herself "conservative" and has little support. Imagine that the "liberal" candidate gets 90% of the popular vote, winning in a "landslide." Suppose that we choose five ballots "at random" out of various ballot boxes scattered "randomly" around the district. (This region uses old-fashioned paper ballots, not electronic voting machines.) What's the probability that we'll select five ballots, all of which indicate votes for the winning candidate? Express the answer as a percentage, rounded off to the nearest whole-number value.

✔ SOLUTION

We must make sure that we define probability correctly here to avoid the "probability fallacy." We seek the proportion of the time that five "randomly" selected ballots will all show votes for the winner, if we select five ballots "at random" on numerous different occasions. For example, if we choose five ballots on 1000 different occasions, we want to know how many of these little "peeks," as a percentage, will produce five ballots all cast for the winner.

Now that we know what we seek, let's convert the winning candidate's percentage to a decimal value. The number 90%, as a proportion, equals 0.9.

We raise this to the fifth power to get the probability P_5 that five "randomly" selected ballots will all show votes for the winner:

$$P_5 = (0.9)^5$$

$$= 0.59049$$

$$= 59.049\%$$

We can round this result off to 59%.

PROBLEM 8-19

Suppose that we pick n ballots "at random" out of various ballot boxes scattered "randomly" around the district in the above scenario, where n represents some arbitrary whole number. What's the probability that all the ballots will indicate votes for the winning candidate?

SOLUTION

Again, we must make sure that we know exactly what we seek. We want to determine the proportion of the time that n "randomly" selected ballots will all show votes for the winner, if we select n ballots "at random" many times.

The probability P_n that n "randomly" selected ballots will all show votes for the winner equals 0.9^n, that is, 0.9 raised to the nth power. Expressed as a percentage, the probability $P_{\%n}$ equals $(100 \times 0.9^n)\%$. The process of calculating this figure can grow tedious for large values of n unless our calculator has an "x to the y power" key. Fortunately, most personal-computer calculator programs, when set for scientific mode, have such a key.

PROBLEM 8-20

Generate a table and plot a graph of $P_{\%n}$ as a function of n, for the scenario described in Problem 8-19. Include values of n from 1 to 10 inclusive.

SOLUTION

See Table 8-8 and Fig. 8-7. In the graph of Fig. 8-7, the dashed curve indicates the general trend, but the values of the function are confined to the individual points.

TABLE 8-8   Table for Problems 8-19 and 8-20. The probability $P_{\%n}$ that n "randomly" selected ballots will all show votes for the winner equals $(100 \times 0.9^n)\%$. We round off the probability figures to the nearest tenth of a percent.

Number of Ballots	Probability That All Ballots Show Winning Vote (%)
1	90.0
2	81.0
3	72.9
4	65.6
5	59.0
6	53.1
7	47.8
8	43.0
9	38.7
10	34.9

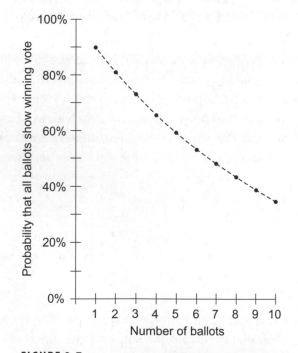

FIGURE 8-7 · Illustration for Problems 8-19 and 8-20.

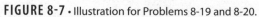

PROBLEM **8-21**

Consider a scenario similar to the one described above, except that the winner receives only 80% of the popular vote. Suppose that we pick n ballots "at random" out of various ballot boxes scattered "randomly" around the district in the above scenario, where n represents some arbitrary whole number. What's the probability that we'll select n ballots, all of which show votes for the winning candidate?

SOLUTION

The probability P_n that n "randomly" selected ballots will all show votes for the winner equals 0.8^n, that is, 0.8 raised to the nth power. Expressed as a percentage, the probability $P_{\%n}$ equals $(100 \times 0.8^n)\%$.

PROBLEM **8-22**

Generate a table and plot a graph of $P_{\%n}$ as a function of n, for the scenario described in Problem 8-21. Include values of n from 1 to 10 inclusive.

SOLUTION

See Table 8-9 and Fig. 8-8. In the graph of Fig. 8-8, the dashed curve indicates the general trend, but the values of the function are confined to the individual points.

TABLE 8-9 Table for Problems 8-21 and 8-22. The probability $P_{\%n}$ that n "randomly" selected ballots will all show votes for the winner equals $(100 \times 0.8^n)\%$. We round off the probability figures to the nearest tenth of a percent.

Number of Ballots	Probability That All Ballots Show Winning Vote (%)
1	80.0
2	64.0
3	51.2
4	41.0
5	32.8
6	26.2
7	21.0
8	16.8
9	13.4
10	10.7

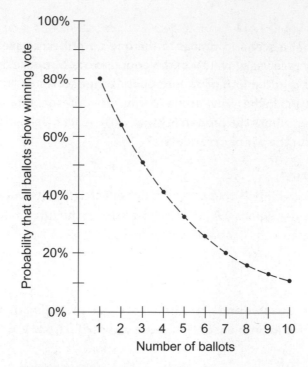

FIGURE 8-8 · Illustration for Problems 8-21 and 8-22.

PROBLEM 8-23

Plot a dual graph showing the scenarios of Problems 8-19 and 8-21 together, thereby comparing the two.

SOLUTION

Figure 8-9 portrays the comparison. The dashed curves indicate the general trends, but the values of the functions are confined to the individual points.

Still Struggling

In the above problems and solutions, we always enclose the expressions "random," "at random," and "randomly" in quotes. We do that because, in theory, doubt exists as to whether or not any human being can carry out a truly random process. These expressions give us some latitude; we can use a pseudorandom number generator or table to derive values on which to base our selections. If we're a little bit lazy, however, we can simply pick out individual people and samples "willy-nilly," pretending that our brains can function in a truly random fashion. If you're still baffled by this business, reread the discussion of randomness at the end of Chap. 7.

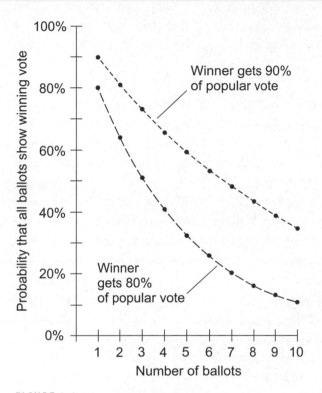

FIGURE 8-9 · Illustration for Problem 8-23.

Data Intervals

The following problems involve quartiles, deciles, percentiles, and straight fractional proportions. Let's consider an example involving climate change. The following scenario constitutes fiction; we use it for illustrative purposes only. It's not real history!

The Distribution

We want to know whether or not the average temperature in the world has increased over the last 100 years. We obtain climate data for many cities and towns scattered throughout the world. We're interested in one figure for each location: the average temperature over the course of last year, versus the average temperature over the course of the year a century ago. The terms "a century ago" and "100 years earlier" means "100 years before this year (99 years before last year)."

In order to calculate a meaningful figure for any given locale, we compare last year's average temperature t, expressed in degrees Celsius (°C), with the average annual temperature s during the course of the year a century ago. We figure the temperature change, T, in degrees Celsius as

$$T = t - s$$

If $t < s$, then T is negative, indicating that the temperature last year was lower than the temperature a century ago. If $t = s$, then $T = 0$, meaning that the temperature last year was the same as the temperature a century ago. If $t > s$, then T is positive, meaning that the temperature last year was higher than the temperature a century ago.

Now imagine that we've obtained data for such a large number of places that we can't generate a table in a reasonable time period or within a reasonable amount of physical space. Instead, we graph the number of locations that have experienced various average temperature changes between last year and a century ago, rounded off to the nearest tenth of a degree Celsius. Suppose that the resulting smoothed-out curve looks like the graph of Fig. 8-10. (We'll deal with the specific case of Thermington in Problems 25 through 30.) We could generate a point-by-point graph made up of many short, straight line segments connecting points separated by 0.1°C on the horizontal scale, but we haven't done that here. Instead, Fig. 8-10 portrays a smooth, continuous graph obtained by curve fitting.

PROBLEM 8-24

What do the points (−2,18) and (+2.8,7) represent in Fig. 8-10?

SOLUTION

These points tell us that there exist 18 locales whose average annual temperatures were lower last year by 2°C as compared with a century ago, as shown by the point (−2,18), and that there exist seven locales whose average annual temperatures were higher last year by 2.8°C as compared with a century ago, as shown by the point (+2.8,7).

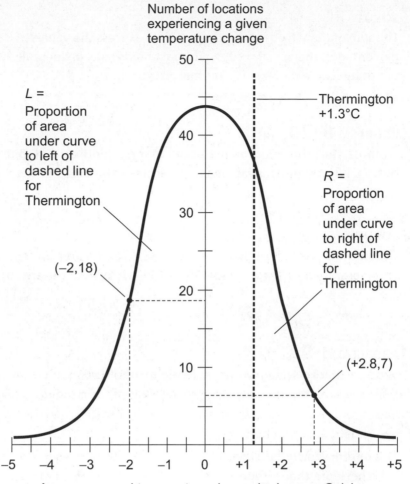

FIGURE 8-10 · Illustration for Problems 8-24 through 8-30.

PROBLEM **8-25**

Suppose that the temperature in the town of Thermington was higher last year by 1.3°C than it was a century ago. We indicate this fact as a heavy, vertical, dashed line in the graph of Fig. 8-10. Suppose that *L* represents the proportion of the area under the curve (but above the horizontal axis showing temperatures) to the left of the vertical dashed line, and *R* represents the proportion of the area under the curve to the right of the vertical dashed line. What can we say about *L* and *R*?

SOLUTION

The sum of L and R equals precisely 1. If we express the values of L and R as percentages, then $L + R = 100\%$. This fact holds true, incidentally, whether or not the curve represents a normal distribution.

PROBLEM 8-26

Suppose that Thermington lies at the 81st percentile point in the distribution. What does this fact mean in terms of the areas of the regions L and R?

SOLUTION

This information tells us that L represents 81% of the area under the curve, and therefore that R represents 100% – 81%, or 19%, of the area under the curve.

PROBLEM 8-27

Suppose that Thermington lies at the eighth decile point in the distribution. What does this fact mean in terms of the areas of the regions L and R?

SOLUTION

This information tells us that L represents 8/10 of the area under the curve, and therefore that R represents 1 – 8/10, or 2/10, of the area under the curve.

PROBLEM 8-28

Suppose that Thermington lies at the third quartile point in the distribution. What does this fact mean in terms of the areas of the regions L and R?

SOLUTION

This information tells us that L represents 3/4 of the area under the curve, and therefore that R represents 1 – 3/4, or 1/4, of the area under the curve.

PROBLEM 8-29

Suppose that Thermington is among the one-quarter of towns in our experiment that experienced the greatest temperature increase between last year and 100 years ago. What does this fact mean in terms of the areas L and R?

SOLUTION

As stated, this problem presents us with an ambiguity. We can interpret it to mean either of two things:

1. We consider all the towns in the experiment
2. We consider only those towns in the experiment that experienced increases in temperature

If we mean (1) above, then the above specification tells us that L represents more than 3/4 of the area under the curve, and therefore that R represents less than 1/4 of the area under the curve. If we mean (2) above, we can say that R represents less than 1/4, or 25%, of the area under the portion of curve that lies to the right of the dependent-variable axis (the vertical axis in the center of graph showing number of locations experiencing a given temperature change); but we can't say anything about L unless we know more about the distribution.

Upon casual observation, the curve in Fig. 8-10 appears symmetrical with respect to the dependent-variable axis. We might want to suppose that the curve constitutes a normal distribution. But no one has assured us of that, and we mustn't assume it as a fact in the absence of proof. Imagine that we run the data through a computer and determine that the curve actually does portray a normal distribution. In that case, if (2) above holds true, L represents more than 7/8 of the total area under the curve, and R represents less than 1/8 of the total area under the curve.

PROBLEM 8-30

Suppose that the curve in Fig. 8-10 represents a normal distribution, and that the dashed vertical line for Thermington lies exactly one standard deviation to the right of the dependent-variable axis. What can we say about the areas L and R in this case?

SOLUTION

Imagine the "sister city" of Thermington, a town called Frigidopolis. Suppose that Frigidopolis was cooler last year, in comparison to 100 years ago, by exactly the same amount that Thermington was warmer. Figure 8-11 shows this situation as a graph. Because Thermington corresponds to a point (or dashed vertical line) exactly one standard deviation to the right of the vertical axis, we can represent Frigidopolis by another point (or dashed vertical line) that lies exactly one standard deviation to the left of the vertical axis.

Based on the appearance of the graphs in Figs. 8-10 and 8-11, and because we know that we have a normal distribution, the mean (μ) coincides with the vertical axis, or the point where the temperature change equals 0. Now let's recall (from Chap. 3) the empirical rule concerning standard deviation (σ) and normal distributions. The vertical dashed line on the left,

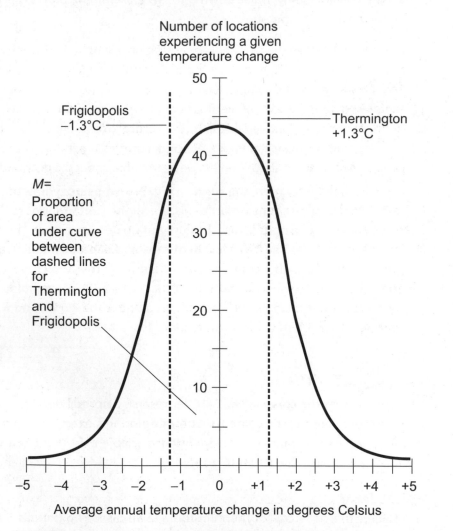

FIGURE 8-11 · Illustration involving the solution to Problem 8-30.

representing Frigidopolis, lies at $-\sigma$ from μ. The vertical dashed line on the right, representing Thermington, lies at $+\sigma$ from μ. The empirical rule tells us that the proportion of the area under the curve between these two dashed lines equals 68% of the total area under the curve, because these two dashed vertical lines represent that portion of the area within $\pm\sigma$ of μ. The proportion of the area under the curve between the vertical axis and either dashed line must therefore equal half this amount, or 34%.

The fact that we're dealing with a normal distribution also tells us that exactly 50% of the total area under the curve lies to the left of μ, and 50% of the total area under the curve lies to the right of μ, as long as we imagine the curve as extending indefinitely to the left or the right. This fact holds true because a normal distribution always exhibits bilateral symmetry with respect to the mean. Knowing all of the foregoing information, we can determine that, relative to the dashed vertical line representing Thermington in Fig. 8-10 on page 283, we have

$$L = 50\% + 34\%$$
$$= 84\%$$

and

$$R = 100\% - 84\%$$
$$= 16\%$$

? Still Struggling

If the graphs of Figs. 8-10 and 8-11 represent normal distributions, you might wonder if any city can exist for which the average annual temperature change attains a truly extreme value, such as a decrease of 80°C or an increase of 120°C in comparison to 100 years ago. In theory, if we had an infinite number of cities on an infinitely large planet from which to choose, we could say yes! Although the curve *approaches* the independent-variable axis (showing temperature change in this scenario), the curve never *actually reaches* the axis, no matter how far out we go. In practice, of course, we have a finite world and a finite number of cities; never in recorded history (let alone in the past century) have temperature extremes of the aforementioned sort occurred.

Sampling and Estimation

Let's do a few review problems involving data sampling and estimation. Recall the steps for conducting a statistical experiment:

- Formulate the question(s) we want to answer.
- Gather sufficient data from the required sources.
- Organize and analyze the data.
- Interpret the information that we've gathered and organized.

PROBLEM 8-31

Most of us have seen charts that tell us how much mass (or weight) our bodies ought to have, in the opinions of medical experts. Medical experts base the ideal mass values on an individual person's height and "frame type" (small, medium, or large), and also on the person's age and gender. Charts made up this year differ from those made up 10 years ago, or 30 years ago, or 60 years ago.

Let's consider hypothetical distributions of human body mass as functions of body height, based on observations of real people rather than on scientific theories. Imagine that we compile such distributions for the whole world and base our data solely on body height, ignoring all other factors. Imagine that we end up with a set of many graphs, one plot for each height in centimeters, ranging from the shortest person in the world to the tallest, showing how massive people of various heights actually are.

Suppose that we obtain a graph (Fig. 8-12, hypothetically) showing the mass distribution of people in the world whose bodies measure 170 centimeters tall, rounded to the nearest centimeter. We round off each individual person's mass to the nearest kilogram. Further suppose that our graph portrays a normal distribution in which the mean, median, and mode all equal 55 kilograms. Of course, we can't put every 170-centimeter-tall person in the world on a scale, measure his or her mass, and record it! Suggest a sampling frame to ensure that Fig. 8-12 constitutes a good representation of the actual distribution of masses for people throughout the world who measure 170 centimeters tall. Suggest another sampling frame that might at first seem workable, but that in fact is not adequate.

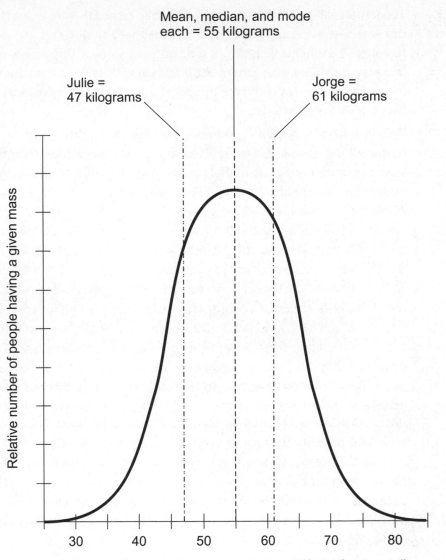

FIGURE 8-12 • Illustration for Problems 8-31 and 8-32.

☑ **SOLUTION** _____

Recall the definition of a sampling frame: a set of elements from within a population, from which we choose a sample. A good sampling frame gives us a representative cross section of a population.

Let's deal with the not-so-good idea first. Suppose that we travel to every recognized country in the world, "randomly" select 100 males and

100 females all 170 centimeters tall, and then record their masses. Then, as our samples, we "randomly" select 10 of these males and 10 of these females. This scheme won't work well, because some countries have much larger populations than others. We'll end up with skewed data that over-represents countries with small populations and underrepresents countries with large populations.

If we modify the preceding scheme, we can get a fairly good sampling frame. We can go to every recognized country in the world and "randomly" select a number of people in that country based on its population in thousands. If a country has, say, 10 million people, we can select, as nearly at random as possible, 10,000 males and 10,000 females all 170 centimeters tall, as our sampling frame. If a country has 100 million people, we can select 100,000 males and 100,000 females all 170 centimeters tall. If a country has only 270,000 people, we can select 270 males and 270 females all 170 centimeters tall. We can have every tenth person stand on a scale and record his or her mass. If a country has so few people that we can't find at least 10 males and 10 females all 170 centimeters tall, we combine that country with one or more others and consider the two together as a single country. From this process, we obtain the distribution by having a computer plot all the individual points and execute a curve-fitting program to generate a smooth graph.

PROBLEM 8-32

Suppose Jorge and Julie both measure 170 centimeters tall. Jorge has a mass of 61 kilograms, and Julie has a mass of 47 kilograms. Where are they in the distribution described above?

SOLUTION

Figure 8-12 shows the locations of Jorge and Julie in the distribution, relative to other characteristics and the curve as a whole.

PROBLEM 8-33

Suppose that we define the "typical human mass" for a person 170 centimeters tall as anything within 3 kilograms either side of 55 kilograms. Also suppose that we define ranges called "low typical" and "high typical," representing masses of more than 3 kilograms but less than 5 kilograms either side of 55 kilograms. Illustrate these ranges in the distribution.

SOLUTION

Figure 8-13 shows these ranges relative to the entire distribution.

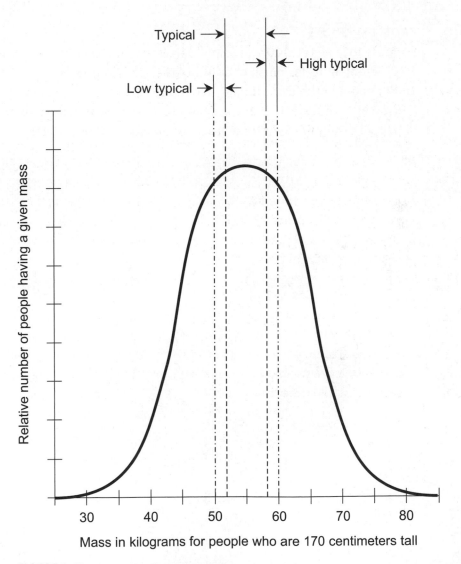

FIGURE 8-13 · Illustration for Problem 8-33.

Hypotheses, Prediction, and Regression

The following several problems involve a fictional but plausible experiment in which we scrutinize the incidence of a mystery illness called Syndrome X. We select 20 groups of 100 people according to various criteria. Then we record the percentage of people in each group who exhibit Syndrome X. Finally, we compile graphs and analyze the data.

PROBLEM 8-34

Suppose that we test 100 "randomly" selected people living at locations at 20 different latitudes north or south of the earth's equator (for a total of 2000 people in the experiment). We render the results as a scatter plot (Fig. 8-14). Latitude acts as the independent variable, and the percentage of people exhibiting Syndrome X acts as the dependent variable.

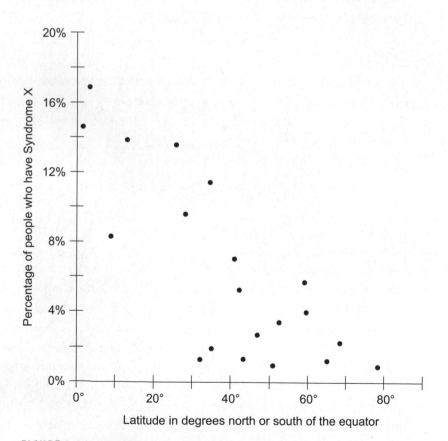

FIGURE 8-14 · Illustration for Problems 8-34 through 8-39.

We come up with a hypothesis that we intend to use as the starting point for our analysis: "If you move closer to, or farther from, the equator, your risk of developing Syndrome X will not change." What sort of hypothesis is this?

Someone says, "Look at Fig. 8-14. Obviously, people who live at low latitudes (i.e., relatively close to the equator) get Syndrome X more often than people who live at high latitudes (far away from the equator). I believe that if you move closer to the equator, your risk of developing Syndrome X will increase." What sort of hypothesis is this?

Someone else says, "The scatter plot shows that a greater proportion of people who live close to the equator have Syndrome X, as compared with people who live far from the equator. But this fact, all by itself, does not *rigorously imply*, according to the rules of formal logic, that if you move closer to the equator, you increase your risk of developing Syndrome X, as compared to what will likely happen if you stay here or move farther from the equator." What sort of hypothesis is this?

SOLUTION

The first hypothesis constitutes a null hypothesis. The second and third hypotheses constitute alternative hypotheses.

PROBLEM 8-35

Provide an argument that supports the null hypothesis described in Problem 34.

SOLUTION

Here's the null hypothesis: "If you move closer to, or farther from, the equator, your risk of developing Syndrome X will not change." The scatter plot of Fig. 8-14 shows that people who live close to the equator have Syndrome X in greater proportion than people who live far from the equator. But we oversimplify matters if we claim that the latitude of residence, all by itself, is responsible for Syndrome X. Suppose, for example, that a person can practically eliminate her chance of getting Syndrome X by taking a medication or precaution that most people who live near the equator don't know about, or in which they don't believe, or that their governments forbid them to take. If you live in England and you know all about

Syndrome X, you might adjust your lifestyle or take a vaccine so that, if you move to Ecuador, you'll bear no greater risk of contracting Syndrome X than you do now.

PROBLEM 8-36

How can we test the null hypothesis described in Problem 34?

SOLUTION

In order to discover whether or not moving from one latitude to another affects the probability that a person will develop Syndrome X, we must test a large number of people who have moved from various specific places to various other specific places. This test will be more complex and time-consuming than the original experiment. Additional factors will enter in, as well. For example, we'll have to find out how long each person has lived in the new location after moving, and how much traveling each person does (e.g., in conjunction with employment). We'll also have to take into account the extent, not only the direction, of the latitude change. Does a difference exist between moving from England to Ecuador, as compared with moving from Sweden to Spain? Another factor is the original residence latitude. Is there a difference between moving from Sweden to Ecuador, as compared with moving from Spain to Ecuador?

? Still Struggling

If the countries mentioned in Solution 8-36 mystify you, find a detailed world map or globe and locate them. Note their general latitudes:

- Ecuador straddles the equator.
- Spain lies a little less than halfway from the equator to the north pole.
- England lies a little more than halfway from the equator to the north pole.
- Sweden lies farther from the equator than any of the other three countries.

PROBLEM 8-37

Figure 8-15 shows a scatter plot of data for the same 20 groups of 100 people that have been researched in our hypothetical survey involving Syndrome X. But instead of the latitude in degrees north or south of the equator, the altitude, in meters above sea level, constitutes the independent variable. What does this graph tell us?

SOLUTION

It's difficult to see any correlation here. Some people might imagine a weak negative correlation between the altitude of a place above sea level and

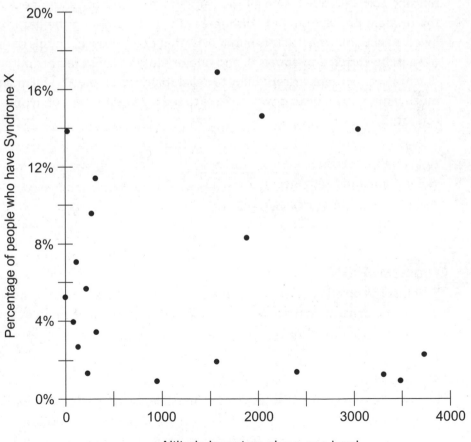

FIGURE 8-15 · Illustration for Problems 8-37 through 8-40.

the proportion of the people exhibiting Syndrome X. Other people might sense a weak positive correlation because of the points in the upper-right portion of the plot. We would have to employ a computer to determine the actual correlation, and when we find it, we should not be surprised if it turns out insignificant.

PROBLEM 8-38

Suppose that we formulate a hypothesis: "If you move to a higher or lower altitude above sea level, your risk of developing Syndrome X does not change." What sort of hypothesis is this?

Someone says, "It seems to me that Fig. 8-15 shows a weak, but not a significant, correlation between altitude and the existence of Syndrome X in the resident population. But I disagree with you concerning the hazards involved with moving. Factors might exist that don't show up in this data, even if the correlation equals 0, and one or more of these factors might drastically affect your susceptibility to developing Syndrome X if you move much higher up or lower down, relative to sea level." What sort of hypothesis is this?

SOLUTION

The first hypothesis constitutes a null hypothesis. The second hypothesis constitutes an alternative hypothesis.

PROBLEM 8-39

Estimate the position of the line of least squares for the scatter plot showing the incidence of Syndrome X versus the latitude north or south of the equator (Fig. 8-14).

SOLUTION

Figure 8-16 shows a "good guess" at the line of least squares for the points in Fig. 8-14.

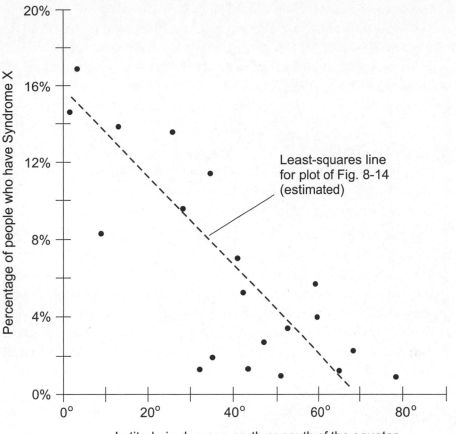

FIGURE 8-16 • Illustration for Problem 8-39.

PROBLEM **8-40**

Figure 8-17 shows a "guess" at a regression curve for the points in Fig. 8-15, based on the notion that a weak negative correlation exists. Does this curve represent a "good guess"? If so, why? If not, why not?

✔ SOLUTION

Figure 8-17 does not represent a "good guess" at a regression curve for the points in Fig. 8-15. We can't make any sort of "good guess" in this situation. The correlation is weak at best, and its nature is uncertain in the absence of computer analysis.

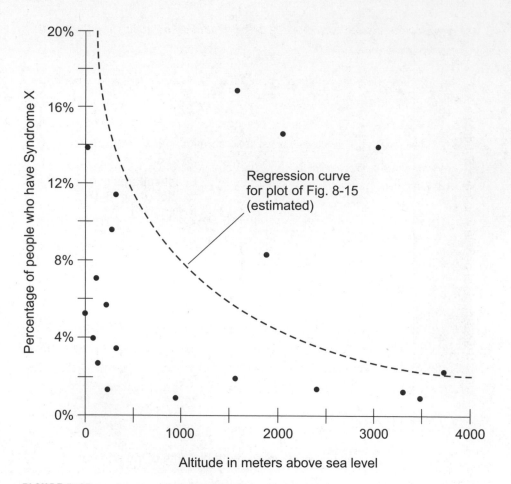

FIGURE 8-17 · Illustration for Problem 8-40.

TIP *The scatter plot of Fig. 8-14 indicates significant correlation, while the plot of Fig. 8-15 indicates insignificant or nonexistent correlation. Now imagine that we encounter a scenario in which the scatter plot shows points less organized than those in Fig. 8-14, but more organized than those in Fig. 8-15. You will wonder, "How can we decide whether or not we should take the correlation, as shown by a scatter plot, seriously?" We can answer this question subjectively in two ways. First, as the number of points in a scatter plot increases (representing more data gathered), we can take the results more seriously. Second, we can render the data, as shown by the points, as ordered pairs and then enter those numbers into a computer programmed to determine correlation, then we can decide upon a* *threshold for how much correlation represents something important. We might, for example, say that:*

- *Any correlation figure between or including −10% and +10% is so small that we don't have to take it seriously*

- *Any correlation figure less than −10% and larger than −30%, or larger than +10% but less than +30%, is substantial enough to take somewhat seriously*

- *Any correlation figure less than or equal to −30%, or larger than or equal to +30%, is substantial enough to take very seriously*

These "spans of correlation" represent entirely subjective notions! You might like them, while your friend considers them worthless.

Correlation and Cause/Effect Relations

As we learned in Chap. 7, the mere existence of a correlation (either positive or negative) between two variables does not logically imply a direct cause/effect relation between the phenomena that the variables represent. Let's solve some hypothetical problems involving correlation and causation.

Two Modern Plagues

During the last few decades of the 20th century and into the first years of the 21st century, a dramatic increase took place in the incidence of adult-onset diabetes (let's call it AOD for short) in the United States of America (USA). In this syndrome, the body has trouble regulating the amount of glucose, an important body fuel, in the blood. During the same period of time, an increase also took place in the percentage of the US population exhibiting obesity. Scientists and physicians have long suspected that cause/effect relationships operate between AOD and obesity, but the exact mechanisms remain subjects for debate.

PROBLEM 8-41

Researchers have found that in the USA, obese adults have AOD in greater proportion than adults who have normal or near-normal weight. If we "randomly" select a person from the USA population, the likelihood of that person having AOD increases as the extent of the obesity increases. What sort of correlation is this?

✔ SOLUTION

Obesity correlates positively with AOD. To state the situation in more technical terms, let M_n represent the normal mass in kilograms for a person of a certain height, gender, and age, and let M_a represent the person's actual mass in kilograms. The probability that a "randomly" selected person in the USA will exhibit AOD is positively correlated with the ratio M_a/M_n.

PROBLEM 8-42

Does the above correlation, all by itself, logically imply that obesity directly causes AOD in the USA? Keep in mind that the term *logical implication*, in this context, means implication in the strongest possible sense. If we say "*P* logically implies *Q*," then in effect we say "If *P*, then *Q*."

✔ SOLUTION

No. The "conventional wisdom" holds that obesity acts as a contributing factor to the development of AOD in the USA, and research has been done to back up this hypothesis. But the mere existence of the correlation, all by itself, does not mathematically prove the hypothesis.

PROBLEM 8-43

Does the above correlation, all by itself, logically imply that AOD directly causes obesity?

✔ SOLUTION

No. Few if any scientists believe that AOD causes obesity in the USA, although some people have suggested that this idea should receive serious attention. The correlation constitutes an observable phenomenon, but if we claim that a direct cause/effect relationship exists between AOD and obesity in the USA, then we must conduct research that demonstrates a plausible explanation to back up that claim.

PROBLEM 8-44

Does the above correlation, all by itself, logically imply that both AOD and obesity in the USA are caused by some third external factor?

✔ SOLUTION

No again! In order to conclude that a cause/effect relationship of any kind exists between or among variables, we must find good reasons to support such a theory. No particular theory logically follows from the mere existence of a correlation.

Most people would agree that at least one causative factor exists for both AOD and obesity in the USA: eating too much food! People who overeat are more likely to suffer from obesity than people who eat only what they need and no more. But even this conclusion is too simple. Some scientists believe that the overconsumption of *certain types of food* correlates more with AOD than the overconsumption of other types of food. Sociological, psychological, and economic issues also come into play. Some people think that the presence of industrial pollutants in water and food correlates positively with the incidence of AOD in the USA. And who knows for sure that the incidence of AOD might not correlate positively with unsuspected factors such as exposure to the electromagnetic fields from ordinary utility power lines?

PROBLEM 8-45

We've seen that the correlation between obesity and AOD in the USA, all by itself, does not *logically imply* any particular cause/effect relationship. Does this fact mean that whenever we see a correlation (however weak or strong, and whether positive or negative), we must conclude that it demonstrates nothing more than a coincidence?

✔ SOLUTION

Once again, the answer is no. When we observe a correlation, we should resist the temptation to come to any specific conclusion about cause/effect in the absence of supporting research or a sound theoretical argument. We should remain skeptical, but we should not close our minds altogether. We should make every possible attempt to determine the truth, without letting personal bias, peer pressure, economic interests, or political pressures cloud our judgment. In many situations, a cause/effect relationship (perhaps more than one!) does exist when we see a correlation.

❓ Still Struggling

Here's an axiom that I like to remember when I look at statistical data or read reports claiming some specific conclusion based on data.

- Let's try to make our theories reflect the truth, rather than trying to make the truth fit our theories.

Let the statistician and the truth-seeker beware!

QUIZ

Refer to the text in this chapter if necessary. A good score is 8 correct. Answers are in the back of the book.

1. Refer to the quiz discussed in Problems 8-1 through 8-15, and whose results we graphed in Fig. 8-6. Imagine that we administer the same quiz to a completely different group of 130 students, resulting in a larger span between $\mu - \sigma$ and $\mu + \sigma$ on the graph of Fig. 8-6. In this case, the standard deviation is

   A. the same as before.
   B. larger than before.
   C. smaller than before.
   D. impossible to determine without more information.

2. Refer again to the quiz discussed in Problems 8-1 through 8-15, and whose results we graphed in Fig. 8-6. Suppose that a controversy arises over one of the questions, causing the professor to discard the question, replace it with a different one, and then recompile all the data after obtaining the students' answers to the replacement question. What effect would this action have on the general shape of the plot?

   A. It would skew the entire plot to the left.
   B. It would skew the entire plot to the right.
   C. It would make the plot appear more sharply peaked.
   D. We need more information before we can say.

3. Refer one last time to the quiz discussed in Problems 8-1 through 8-15, and whose results we graphed in Fig. 8-6. Suppose that we administer another 10-question test to the same group of students and get the following results:

   - Eight people missed all the questions
   - Six people got one correct answer
   - Eight people got two correct answers
   - Eleven people got three correct answers
   - Eighteen people got four correct answers
   - Nineteen people got five correct answers
   - Nineteen people got six correct answers
   - Twelve people got seven correct answers
   - Twelve people got eight correct answers
   - Ten people got nine correct answers
   - Seven people wrote perfect papers

   What effect will these new results have on the general shape of the plot when we create another graph relating the same two variables as in Fig. 8-6?

   A. It will not have any effect.
   B. It will yield a more sharply peaked plot.
   C. It will yield a less sharply peaked plot.
   D. We need more information before we can say.

TABLE 8-10	Table for Quiz Question 4.	
Test Score	Absolute Frequency	Cumulative Absolute Frequency
0	8	8
1	6	14
2	8	22
3	11	33
4	16	49
5	21	70
6	20	90
7	8	98
8	8	106
9	15	121
10	9	130

4. Table 8-10 shows the results of a hypothetical quiz given to a class of 130 students. What, if any, serious errors or anomalies does this table have?

   A. We can't calculate the mean.
   B. We can't determine the median score.
   C. We can't determine any mode score.
   D. No serious errors or anomalies exist in Table 8-10.

5. What's the probability that an unbiased dodecahedral (12-faced) die will, if tossed five times in a row, come up showing face number 1 on all five occasions?

   A. 1 in 5040
   B. 1 in 248,832
   C. 1 in 244,140,625
   D. 1 in 479,001,600

6. Imagine that we perform further research concerning adult-onset diabetes (AOD) as described earlier in this chapter. We notice that in the general population throughout the world, people with AOD have skin problems more often than people without AOD. This data logically implies that

   A. skin problems cause AOD.
   B. AOD causes skin problems.
   C. an unknown factor causes both skin problems and AOD.
   D. skin problems correlate positively with AOD.

7. **Refer again to the Syndrome X scenario discussed in Problems 8-34 through 8-40. Figure 8-18 is a scatter plot showing the incidence of Syndrome X versus the percentage of land area covered by forest at the location where each person lives. This plot indicates**

    A. insignificant or zero correlation between the incidence of Syndrome X and the percentage of land area covered by forest.

    B. positive correlation between the incidence of Syndrome X and the percentage of land area covered by forest.

    C. negative correlation between the incidence of Syndrome X and the percentage of land area covered by forest.

    D. a cause/effect relationship between the incidence of Syndrome X and the percentage of land area covered by forest.

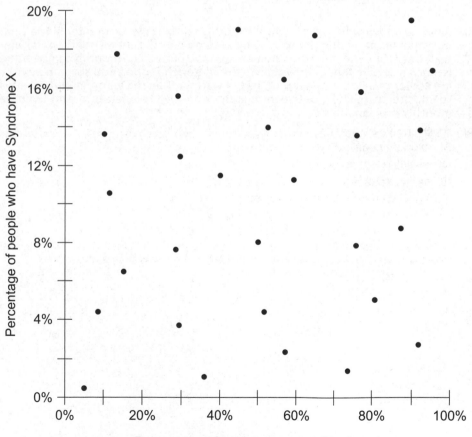

FIGURE 8-18 · Illustration for Quiz Questions 7 and 8.

8. Figure 8-18 shows no least-squares line. If we use a computer in an attempt to determine the least-squares line for this scatter plot, the computer will likely yield

    A. a horizontal line.

    B. a vertical line.

    C. a line that ramps downward as we move toward the right.

    D. no line at all or an "error" message.

9. Which of the following statements is false?

    A. The mean, median, and mode might all differ in a statistical distribution.

    B. In a normal distribution, the variance equals the square of the standard deviation.

    C. The graph of a normal distribution never appears symmetrical with respect to the mean.

    D. The existence of a correlation between two phenomena does not necessarily indicate a cause/effect relationship.

10. Imagine a town where the average monthly rainfall is *much* greater in the summer than in the winter; the spring and autumn months are wetter than the winter months but drier than the summer months. The average monthly temperature is *much* higher in the summer than in the winter; the spring and autumn months are cooler than the summer months but warmer than the winter months. In this town, the correlation between average monthly temperature and average monthly precipitation is

    A. strongly positive.

    B. strongly negative.

    C. weakly positive.

    D. weakly negative.

Test: Part II

Do not refer to the text when taking this test. You may draw diagrams or use a calculator if necessary. A good score is at least 45 correct. Answers are in the back of the book. It's best to have a friend check your score the first time, so you won't memorize the answers if you want to take the test again.

1. Consider a scatter plot between an independent variable x and a dependent variable y, both of which are confined to positive real-number values. Imagine that the points in the plot all lie on or near a curve with the equation $y = x^3$. What's the general position or orientation of the least-squares line?

   A. It's horizontal with positive values.
   B. It's horizontal with negative values.
   C. It ramps upward as we move toward the right.
   D. It ramps downward as we move toward the right.
   E. We need more information to answer this question.

2. Suppose that you flip a two-sided, flat coin seven times. What's the probability that the coin will come up "tails" on all seven tosses? Assume the coin is not "weighted," so that the probability of it coming up "heads" on any given flip equals 50%.

   A. 1 in 49
   B. 1 in 128
   C. 1 in 2401
   D. 1 in 4096
   E. 1 in 16,384

3. Suppose that you flip a tetrahedral die (a die with the shape of a four-faced regular polyhedron) seven times onto a flat, horizontal table. What's the probability that the die will land with the same face against the table on all seven tosses? Assume the die is not "weighted," so that the probability of it landing with any particular face against the table on any given toss equals 25%.

   A. 1 in 49
   B. 1 in 128
   C. 1 in 2401
   D. 1 in 4096
   E. 1 in 16,384

4. Imagine that the points in a scatter plot all lie fairly near a line that ramps downward as we move toward the right in the coordinate grid. The values along the horizontal axis increase as we move toward the right. The values along the vertical axis increase as we move upward. What's the correlation between the variables in this situation?

   A. 0
   B. Something between 0 and +1
   C. +1
   D. Something between −1 and 0
   E. −1

5. Imagine that you want to paint the exterior of your house. You estimate that it will take you 20 hours to complete the job. Your brother says it will take you more than 20 hours. Your sister says will take you less than 20 hours. Your own estimate of the required time constitutes

   A. the null hypothesis.
   B. a one-sided alternative hypothesis.

C. a two-sided alternative hypothesis.
D. an unprovable proposition.
E. a regressive proposition.

6. **Imagine that we sample an element of a set, and then we leave the element in the set so that we can sample it again. In this situation, we perform sampling with**

A. independence.
B. replacement.
C. redundancy.
D. dependence.
E. ambiguity.

7. *Inference* **is the technical term that we use to describe the process of**

A. formulating a null hypothesis.
B. formulating alternative hypotheses.
C. sampling data from a population.
D. deriving conclusions from data and hypotheses.
E. calculating the correlation between two variables.

8. **A regression curve for the points in a scatter plot *always* takes the form of a**

A. circle.
B. parabola.
C. straight line.
D. hyperbola.
E. None of the above

9. **Figure Test II-1 is a generic graph of a normal distribution showing the probability of a hypothetical "event X" as a function of an independent variable. What's the approximate ratio of the shaded area to the total area between the curve and the independent-variable axis?**

A. 0.35.
B. 0.50.
C. 0.68.
D. 0.71.
E. We need more information to answer this question.

10. **In Fig. Test II-1, what do we call the range of values marked "Interval" between the leftmost and rightmost dashed vertical lines?**

A. A normal interval
B. A deviation interval
C. A confidence interval
D. A population interval
E. A sampling interval

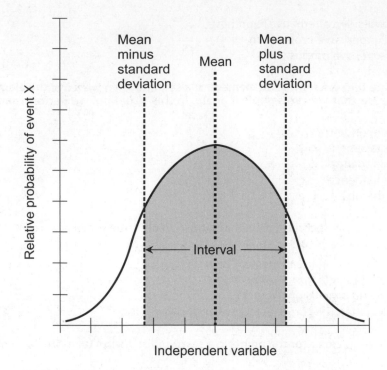

FIGURE TEST II-1 · Illustration for Part II Test Questions 9 and 10.

11. **The term "butterfly effect" refers to a situation in which**
    A. a small event has major consequences.
    B. events occur in bunches, rather than evenly spread out over time.
    C. a major event has multiple, or even infinitely many, small consequences.
    D. confidence intervals fluctuate unpredictably as time passes.
    E. attempts to generate "truly random" numbers yield unintended results.

12. **What's the *widest* possible range of correlation values *r* that we can observe between any two variables?**
    A. $0 \leq r < +1$
    B. $-1/2 < r < +1/2$
    C. $-1 < r \leq 0$
    D. $-0.68 < r < +0.68$
    E. None of the above

13. **Fill in the blank in the following sentence to make it true: "All the sampling frames we choose during a statistical experiment must constitute _____ of the population."**
    A. the largest possible proportions
    B. unbiased representations

C. deciles, quartiles, or percentiles

D. confidence intervals

E. equal proportions

14. **In a scatter plot showing a moderate to strong positive or negative correlation, the outliers, if any, appear**

A. near the axes.

B. near the regression curve.

C. scattered "at random."

D. far from the least-squares line.

E. near the origin.

15. **Imagine a gigantic population *P* in which some characteristic can vary from element to element. We choose a large number of samples from *P*. Each sample represents a different "random" cross section of *P*, but all the samples have equal size. We find the mean of each sample and graph the results to obtain a sampling distribution of means. According to the Central Limit Theorem, the sampling distribution of means constitutes a normal distribution**

A. if the distribution for *P* is normal.

B. if the distribution for *P* appears as a smooth curve.

C. if it contains uniformly distributed outliers.

D. if the regression curve appears as a straight line.

E. under no circumstances.

16. **Refer to Table Test II-1. This shows the results of a hypothetical 10-question test given to a class of students. What's the mode?**

A. 5.00

B. 5.50

C. 5.67

D. 6.00

E. 6.33

17. **What's the median in Table Test II-1?**

A. 5.00

B. 5.50

C. 5.67

D. 6.00

E. 6.33

TABLE TEST II-1   Table for Part II Test Questions 16 through 18.

Test Score	0	1	2	3	4	5	6	7	8	9	10
Absolute Frequency	0	4	6	20	25	40	42	32	28	10	2

18. **What's the mean in Table Test II-1?**

    A. 5.00
    B. 5.50
    C. 5.67
    D. 6.00
    E. 6.33

19. **In a normal distribution, suppose that we consider a range of values between the mean minus k standard deviations and the mean plus k standard deviations, where k represents a positive real-number constant. We can define that span as a**

    A. normal interval.
    B. deviation interval.
    C. confidence interval.
    D. population interval.
    E. sampling interval.

20. **An alternative hypothesis always constitutes a**

    A. specific assumption other than the null hypothesis.
    B. statement that we can't prove true.
    C. statement that we can prove true.
    D. statement that gives rise to a contradiction.
    E. conclusion we reach only after exhausting all other possibilities.

21. **Imagine that a typical person in the town of Fitsville gets an average of between 40 and 60 minutes of exercise every day. This statement is equivalent to saying that a typical person in Fitsville gets**

    A. 50 minutes of exercise a day, plus or minus 5%.
    B. 50 minutes of exercise a day, plus or minus 10%.
    C. 40 minutes of exercise a day, plus up to 20%.
    D. 1 hour of exercise a day, minus up to 20%.
    E. None of the above

22. **In Fig. Test II-2, the dashed line represents an example of a**

    A. least-squares curve.
    B. regression curve.
    C. correlation-mean curve.
    D. scattering curve.
    E. bimodal curve.

23. **In the situation shown by Fig. Test II-2, the correlation between the values of the variables appears to be**

    A. strongly negative.
    B. weakly negative.
    C. zero.
    D. weakly positive.
    E. strongly positive.

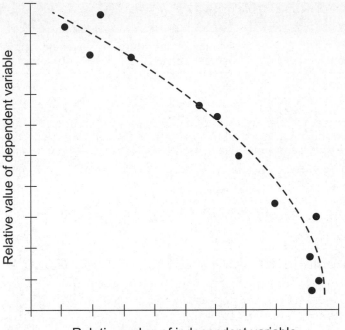

Relative value of independent variable

FIGURE TEST II-2 • Illustration for Part II Test Questions 22 and 23.

24. **Which of the following methods A, B, C, or D represents the best way to generate random digits?**
   A. List the digits (0 through 9) one by one after the decimal point in an irrational number such as the square root of 17.
   B. Operate a "lottery machine" in which air jets blow a set of 10 lightweight balls, numbered 0 through 9, around inside a pressurized bottle.
   C. Write down individual decimal digits (0 through 9) on a sheet of paper, in as nearly a "random" fashion as we can manage.
   D. Choose a number and then double it over and over, writing down the digits as they appear sequentially in the products.
   E. All of the preceding four methods represent equally good ways to generate random digits.

25. **Table Test II-2 illustrates the average monthly temperatures and rainfall amounts for an imaginary city in the Southern Hemisphere called New Dublinsburg. From this table, it appears that**
   A. no correlation exists between the average monthly temperature and the average monthly rainfall.
   B. a positive correlation exists between the average monthly temperature and the average monthly rainfall.

Month	Average Temperature, Degrees celsius	Average Rainfall, Centimeters
TABLE TEST II-2 Table for Part II Test Questions 25 and 26.		
January	22.0	5.1
February	22.5	6.0
March	19.6	8.5
April	17.0	10.5
May	15.7	12.7
June	13.1	14.0
July	11.0	16.5
August	11.5	17.0
September	12.9	12.5
October	14.6	6.0
November	18.0	4.7
December	20.1	4.3

C. a negative correlation exists between the average monthly temperature and the average monthly rainfall.

D. a correlation exists, but we need more information to know whether it's positive or negative.

E. the data in the table represent an impossible scenario that we can discard as completely meaningless.

26. Refer to Table Test II-2 and Fig. Test II-3. Which of the graphs in Fig. Test II-3 most nearly represents the line of least squares for a scatter plot of the relationship between the average monthly temperature and the average monthly rainfall for New Dublinsburg? Assume that on each plot, temperatures increase as you move to the right along the horizontal scale, and rainfall amounts increase as you move up along the vertical scale.

A. Graph A

B. Graph B

C. Graph C

D. Graph D

E. None of the above

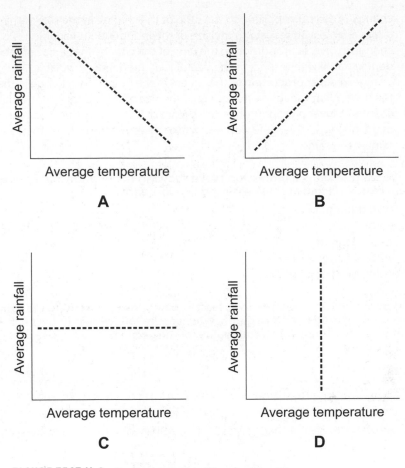

FIGURE TEST II-3 · Illustration for Part II Test Question 26.

27. **In order to determine whether or not a given null hypothesis is correct, we must**

    A. formulate multiple alternative hypotheses.
    B. subject it to a "common-sense" test.
    C. test it in a computer program designed to predict random outcomes.
    D. conduct an experiment to find out for sure.
    E. rigorously prove it using the rules of mathematical logic.

28. **In a situation where two variables correlate, suppose that we alter the *size but not the essence* of the measurement unit of either or both variables. For example, we might change a unit from inches to centimeters (both expressing length), but not from inches to kilograms (one expressing length and the other expressing mass). In any such case, the quantitative correlation, *r*, between the two variables will**

A. change in direct proportion to the ratio of the change in the size of the dependent variable with respect to the size of the independent variable.

B. change in inverse proportion to the ratio of the change in the size of the dependent variable with respect to the size of the independent variable.

C. change in direct proportion to the square of the ratio of the change in the size of the dependent variable with respect to the size of the independent variable.

D. change in inverse proportion to the square of the ratio of the change in the size of the dependent variable with respect to the size of the independent variable.

E. remain the same.

29. **If we want to draw a regression curve showing the relationship between two variables, we'll find our task easiest if we work with a**

   A. paired-bar graph.
   B. scatter plot.
   C. Venn diagram.
   D. vertical bar graph.
   E. horizontal bar graph.

30. **Suppose that we've determined a least-squares line representing the correlation between two variables on a graph. Which of the following numbers A, B, C, or D, if any, represents an *implausible* value for the *geometric slope* ("rise over run") of the least-squares line?**

   A. −2.00
   B. −1.00
   C. +0.25
   D. +1.00
   E. All of the above numbers represent plausible geometric slope values for a least-squares line.

31. **Suppose that we've determined a least-squares line representing the correlation between two variables on a scatter plot. Which of the following numbers A, B, C, or D, if any, represents an *implausible* value for the *correlation* between the variables?**

   A. +2.00
   B. +0.98
   C. +0.25
   D. −0.76
   E. All of the above numbers represent plausible values for the correlation between two variables.

32. **Fill in the blank to make the following sentence correct: "In a scatter plot, we can illustrate the general way in which the two variables relate by drawing a _____ curve for the points."**

   A. null
   B. parabolic
   C. probability

    D. regression
    E. normal

33. **In a vertical bar graph, what (if any) advantage do we gain by numerically show-ing the dependent-variable values at the tops of the bars?**

    A. It makes the graph appear less cluttered to observers.
    B. It eliminates graphical observation errors.
    C. It allows observers to infer the least-squares line.
    D. It allows observers to infer the regression curve.
    E. We gain no advantage by numerically showing the dependent-variable values at the tops of the bars.

34. **With respect to a statistician's source data, the term *primary* means that**

    A. the data contains extra or irrelevant points or values.
    B. some of the data points or values are missing.
    C. the data has already been obtained by someone else.
    D. the statistician has directly gathered all of the data.
    E. the statistician has interpolated missing data points or values.

35. **When we perform calculations to determine a large confidence interval (many estimates of the standard deviation either side of the estimate of the mean), we must ensure that**

    A. the slope of the least-squares line lies between −1 and 1.
    B. we consider a range of values that encompasses the entire distribution.
    C. we work only with primary source data.
    D. we have a clear graph showing the regression curve for all the data points.
    E. the range of values is a small fraction of the estimate of the mean.

36. **Suppose that a scatter plot shows zero correlation between two variables. How many least-squares lines can exist for this plot?**

    A. None
    B. One
    C. Two
    D. As many as we want
    E. Infinitely many

37. **When we portray two variable quantities side by side in a vertical bar graph, our graph illustrates**

    A. concomitant data.
    B. dependent data.
    C. regressive data.
    D. inverse data.
    E. paired data.

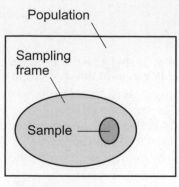

FIGURE TEST II-4 · Illustration for
Part II Test Questions 38 and 39.

38. **Figure Test II-4 is a**

   A. scatter plot.
   B. sampling diagram.
   C. population plot.
   D. Venn diagram.
   E. framing plot.

39. **Figure Test II-4 portrays the fact that**

   A. a population constitutes a subset of a sampling frame, which in turn constitutes a subset of a sample.
   B. a sample constitutes a subset of a sampling frame, which in turn constitutes a subset of a population.
   C. any element of a given population must also constitute an element of every sampling frame within it.
   D. any element of a given sampling frame must also constitute an element of every sample within it.
   E. a sampling frame is always contained within a sample, which in turn is always contained within a population.

40. **Figure Test II-5 is a**

   A. scatter plot.
   B. sampling diagram.
   C. population plot.
   D. Venn diagram.
   E. framing plot.

41. **How is the least-squares line oriented for the data points shown as small black dots (and therefore *not* including *P*, *Q*, or *R*) in Fig. Test II-5?**

   A. It runs at a slant from the lower left corner of the coordinate grid to the upper right corner.
   B. It runs at a slant from the upper left corner of the coordinate grid to the lower right corner.

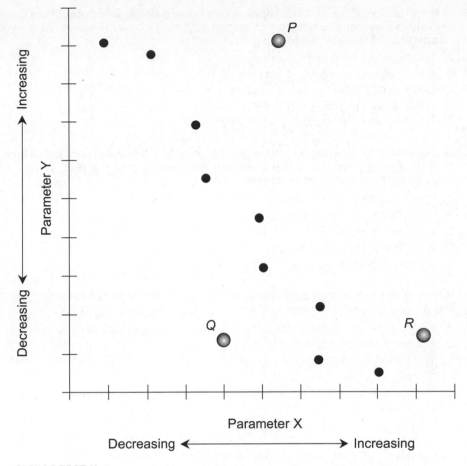

FIGURE TEST II-5 · Illustration for Part II Test Questions 40 through 45.

C. It runs vertically from the bottom center of the coordinate grid to the top center.

D. It runs horizontally from the left middle of the coordinate grid to the right middle.

E. It's not a straight line at all, but a parabolic arc.

42. **In the situation of Fig. Test II-5, suppose that we add a new data point in the location shown by point P without changing anything else in the plot. What will happen to the slope of the least-squares line?**

A. It will change from negative to positive.

B. It will remain negative, but get steeper.

C. It will remain positive, but get steeper.

D. It will remain negative, but get less steep.

E. It will remain positive, but get less steep.

43. In the situation of Fig. Test II-5, suppose that we add a new data point in the location shown by point *Q* without changing anything else in the plot. What will happen to the slope of the least-squares line?

    A. It will change from negative to positive.
    B. It will remain negative, but get steeper.
    C. It will remain positive, but get steeper.
    D. It will remain negative, but get less steep.
    E. It will remain positive, but get less steep.

44. In the situation of Fig. Test II-5, suppose that we add a new data point in the location shown by point *R* without changing anything else in the plot. What will happen to the slope of the least-squares line?

    A. It will change from negative to positive.
    B. It will remain negative, but get steeper.
    C. It will remain positive, but get steeper.
    D. It will remain negative, but get less steep.
    E. It will remain positive, but get less steep.

45. Which of the following statements A, B, C, or D, if any, is *false* concerning graphical diagrams such as the one shown in Fig. Test II-5?

    A. Negative correlation translates to a least-squares line with negative slope.
    B. Positive correlation translates to a least-squares line with positive slope.
    C. The correlation figure equals the geometric slope ("rise over run") of the least-squares line.
    D. The correlation figure tells us, in general terms, how nearly the data points lie to the least-squares line.
    E. All of the above statements are true.

46. Chaos theory offers an explanation for why

    A. events tend to occur in bunches.
    B. events tend to average out over time.
    C. scatter plots tend to get less orderly over time.
    D. correlation does not logically imply causation.
    E. causation does not logically imply correlation.

47. What's the technical term for the type of graph shown in Fig. Test II-6?

    A. Paired-bar graph
    B. Relational graph
    C. Scatter plot
    D. Correlation plot
    E. Regression graph

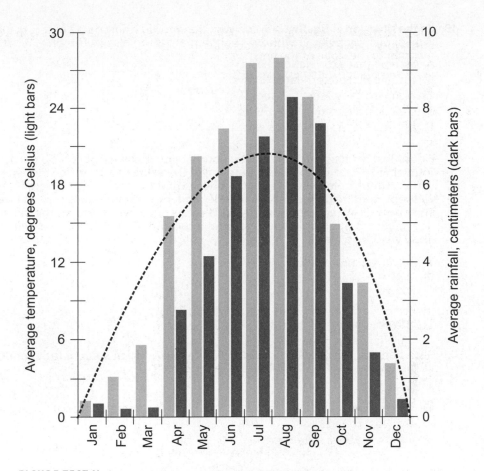

FIGURE TEST II-6 · Illustration for Part II Test Questions 47 through 49.

48. **What can we say about the correlation *r* between the monthly average tempera-ture and rainfall data in the situation portrayed by Fig. Test II-6?**

   A. $-1 < r < 0$

   B. $0 < r < +1$

   C. $r = -1$

   D. $r = +1$

   E. $r = 0$

49. **Suppose that we use a computer in an attempt to locate the least-squares line in Fig. Test II-6. We should expect the computer to come up with**

   A. a line that ramps downward as we move toward the right.

   B. a line that ramps upward as we move toward the right.

   C. a curve that coincides (more or less) with the dashed curve in the figure.

   D. a horizontal line.

   E. nothing, because least-squares lines don't apply to this type of graph.

50. Suppose that we create, and then deliberately reject, a null hypothesis in a certain two-variable scenario. In effect we

    A. assume that a correlation exists.

    B. assume that no correlation exists.

    C. assume a cause/effect relationship.

    D. create an alternative hypothesis.

    E. assume that no hypothesis exists.

51. Figure Test II-7 illustrates the relative probability that we'll observe a particular "outcome X" (vertical axis) as a function of the value of a specific independent variable (horizontal axis). We conduct five experiments over five different intervals representing defined spans of independent-variable values (vertical bars). The intervals all represent equally wide spans of values. Based on the information shown in this graph, which span of independent-variable values will most likely yield "outcome X"?

    A. Span A

    B. Span B

    C. Span C

    D. Span D

    E. Span E

52. Based on the information shown in Fig. Test II-7, which span of independent-variable values will least likely yield "outcome X"?

    A. Span A

    B. Span B

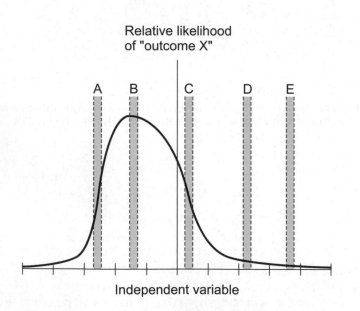

FIGURE TEST II-7 · Illustration for Part II Test Questions 51 through 53.

C. Span C

D. Span D

E. Span E

53. **Based on the information shown in Fig. Test II-7, which two spans of independent-variable values yield approximately equal chances of "outcome X"?**

A. Spans A and B

B. Spans A and C

C. Spans B and C

D. Spans C and E

E. None of the above

54. **Imagine that we draw a straight line somewhere "at random" on the coordinate grid of a 100-point scatter plot. We measure the distance between the line and each point, one by one, using the same units every time. This process yields a set of 100 distance values s_1 through s_{100}. We employ a computer to calculate the sum**

$$S = s_1^2 + s_2^2 + s_3^2 + \ldots + s_{100}^2$$

Next, we use our computer in an attempt to locate a unique, single straight line in the scatter plot that produces the smallest possible value for S. Our computer display gives us a vivid, animated show as it works out the problem. Finally, the machine comes up with a clear result and portrays the sought-after line on the coordinate grid. We have found the scatter plot's

A. regression line.

B. normal line.

C. correlation line.

D. least-squares line.

E. implication line.

55. **Suppose that, in the scenario of Question 54, our computer fails to locate any single line for which the value of S attains a minimum. The program generates an error signal or crashes the computer. Based on this situation, we can suspect that the correlation between the variables portrayed on the scatter plot's axis is**

A. undefined.

B. "negative infinity."

C. equal to 0.

D. equal to −1.

E. equal to some value less than −1 or greater than +1.

56. **When we gather data, we must make sure that each sampling frame**

A. includes every element of the population.

B. contains every element of the sample.

C. constitutes an unbiased representation of the population.

D. correlates positively with all the other sampling frames.

E. exhibits zero correlation with respect to the population.

57. **If we conduct a statistical experiment with a set that contains *infinitely many* elements, then the size of that set**

    A. increases if we perform sampling with replacement, but remains constant if we perform sampling without replacement.
    B. decreases if we perform sampling with replacement, but remains constant if we perform sampling without replacement.
    C. remains constant if we perform sampling with replacement, but increases if we perform sampling without replacement.
    D. remains constant if we perform sampling with replacement, but decreases if we perform sampling without replacement.
    E. remains constant whether we perform sampling with or without replacement.

58. **If we conduct a statistical experiment with a set that contains a *large but finite* number of elements, then the size of that set**

    A. increases if we perform sampling with replacement, but remains constant if we perform sampling without replacement.
    B. decreases if we perform sampling with replacement, but remains constant if we perform sampling without replacement.
    C. remains constant if we perform sampling with replacement, but increases if we perform sampling without replacement.
    D. remains constant if we perform sampling with replacement, but decreases if we perform sampling without replacement.
    E. remains constant whether we perform sampling with or without replacement.

59. **What's the *narrowest* possible range of values that we can ever expect to see for the correlation *r* between two variables?**

    A. $-0.00068 < r < +0.00068$
    B. $-0.0068 < r < +0.0068$
    C. $-0.068 < r < +0.068$
    D. $-0.68 < r < +0.68$
    E. None of the above

60. **We would find a curve-fitting computer program especially useful for graphically illustrating**

    A. the regression for a scatter plot.
    B. the correlation in a bar graph.
    C. cause/effect relationships among variables.
    D. populations within a sampling frame.
    E. sampling without replacement in a finite population.

Final Exam

Do not refer to the text when taking this exam. You may draw diagrams or use a calculator if necessary. A good score is at least 75 correct. You'll find the answers listed in the back of the book. Have a friend check your score, so you won't memorize the answers if you want to take the test again.

1. **If a distribution contains an odd number of elements, then we define the median as the**

   A. sum of all the values, divided by the square of the number of values.

   B. mathematical average of all the values, multiplied by the square root of the number of values.

   C. value such that the number of elements greater than or equal to it is the same as the number of elements less than or equal to it.

   D. value such that the number of elements greater than or equal to it exceeds the number of elements less than or equal to it.

   E. value such that the number of elements less than or equal to it exceeds the number of elements greater than or equal to it.

2. **When two values "compete" for a calculated median, we define the actual median as**

   A. the square root of their product.

   B. their mathematical average.

   C. the smaller value.

   D. the larger value.

   E. nothing at all; instead, we say that the distribution has no median.

3. **Fill in the blank to make the following sentence true: "If we conduct an experiment on a sample set that contains infinitely many elements, that set _____ whether we conduct replacement or not."**

   A. can't be defined

   B. eventually becomes finite

   C. gradually increases in size

   D. gradually decreases in size

   E. maintains its size

4. **The *intersection* of the set of irrational numbers with the set of rational numbers equals the**

   A. set of integers.

   B. set of irrational numbers.

   C. set of rational numbers.

   D. set of real numbers.

   E. empty set.

5. The *union* of the set of irrational numbers with the set of rational numbers equals the

   A. set of integers.

   B. set of irrational numbers.

   C. set of rational numbers.

   D. set of real numbers.

   E. empty set.

6. In the course of a statistical experiment, we collect data, organize the data into information, and finally interpret the information to obtain

   A. samples.

   B. confidence.

   C. populations.

   D. knowledge.

   E. opinions.

7. How does the set N of natural numbers compare with the set Z of integers?

   A. $N \in Z$

   B. $N \subseteq Z$

   C. $Z \in N$

   D. $Z \subseteq N$

   E. $Z = N$

8. As we increase the size of a sampling frame compared to the size of a population, what should we expect?

   A. It will constitute a more accurate representation of the population.

   B. It will become more "random."

   C. It will become less "random."

   D. It will constitute a less accurate representation of the population.

   E. We will have to expend less time, energy, and money to evaluate it.

9. Which of the following statements A, B, C, or D, if any, can *never* hold true concerning a hypothesis?

   A. It turns out correct.

   B. It turns out wrong.

C. We never find out whether it turns out correct or not.

D. We formulate it as an "educated guess."

E. None of the above; that is, any of the foregoing conditions can hold true concerning a hypothesis.

10. Figure Exam-1 illustrates the price of a hypothetical stock versus the time, on a continuous basis, during a particular day from 10:00 a.m. until 3:00 p.m. This figure is an example of a

A. histogram.

B. point-to-point graph.

C. scatter plot.

D. normal distribution.

E. None of the above

11. In the graph of Fig. Exam-1, the vertical-line test implies that, during the time period shown,

A. the stock price constitutes a function of the time.

B. the time constitutes a function of the stock price.

C. the stock price constitutes a function of the time, and vice versa.

D. neither variable constitutes a function of the other variable.

E. things can't possibly have taken place this way.

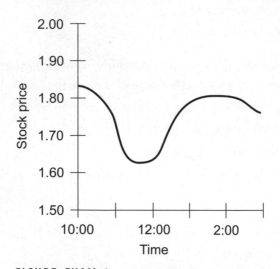

FIGURE EXAM-1 • Illustration for Final Exam Questions 10 and 11.

12. When we use a computer to generate "artificial reality" to simulate a "real-world" scenario, we call our creation

    A. an experiment.

    B. a null hypothesis.

    C. a model.

    D. a probability range.

    E. an alternative hypothesis.

13. We can loosely define the term *mode* for a discrete variable in a statistical distribution as the

    A. value(s) that occur(s) with the highest absolute frequency.

    B. value(s) that occur(s) with the lowest absolute frequency.

    C. sum of all the values, divided by the number of values.

    D. square root of the product of all the values.

    E. point representing the "middle" of the distribution.

14. In a normal distribution, we define the ratio of the standard deviation to the mean as the

    A. Z score.

    B. coefficient of variation.

    C. variance.

    D. interquartile range.

    E. central tendency.

15. Fill in the blank to make the following sentence true: "In a _____ distribution, the function remains constant for all values of the random variable within a defined span."

    A. regressive

    B. uniform

    C. discrete

    D. normal

    E. linear

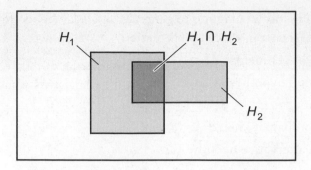

FIGURE EXAM-2 • Illustration for Final Exam Question 16.

16. In the Venn diagram of Fig. Exam-2, the dark-shaded region portrays the intersection of the two sample spaces H_1 and H_2 (light-shaded regions) for an arbitrarily large set of identical experiments in which we observe outcomes h_1 and h_2. Suppose that

$$p(h_1 \wedge h_2) = p(h_1)\, p(h_2)$$

where h_1 and h_2 represent the outcomes, and the inverted V symbol ($\wedge$) stands for the logical connector "and." Based on this information, we can conclude that the outcomes are

A. independent.

B. complementary.

C. supplementary.

D. contradictory.

E. mutually exclusive.

17. Where does the 33rd percentile point lie in the data set shown by Table Exam-1?

A. Exactly at the score of 19

B. Between the scores of 19 and 20

C. Exactly at the score of 29

D. Between the scores of 29 and 30

E. Exactly at the score of 30

TABLE EXAM-1	Results of a hypothetical 40-question test taken by 1000 students. This table goes with Final Exam Questions 17 through 19.	
Test Score	Absolute Frequency	Cumulative Absolute Frequency
0	5	5
1	5	10
2	10	20
3	14	34
4	16	50
5	16	66
6	18	84
7	16	100
8	12	112
9	17	129
10	16	145
11	16	161
12	17	178
13	22	200
14	13	213
15	19	232
16	18	250
17	25	275
18	25	300
19	27	327
20	33	360
21	40	400
22	35	435

(*Continued*)

TABLE EXAM-1 (Continued)		
Test Score	Absolute Frequency	Cumulative Absolute Frequency
23	30	465
24	35	500
25	31	531
26	34	565
27	35	600
28	34	634
29	33	667
30	33	700
31	50	750
32	50	800
33	45	845
34	27	872
35	28	900
36	30	930
37	28	958
38	20	978
39	12	990
40	10	1000

18. Suppose that you're one of the 1000 students who took the test whose results appear in Table Exam-1. If you got 28 correct answers, in which 25% range have you scored?

A. The zeroth

B. The first

C. The second

D. The third

E. The fourth

19. Suppose that you're one of the 1000 students who took the test whose results appear in Table Exam-1. If you got 22 correct answers, in which 10% range have you scored?

    A. The second

    B. The fourth

    C. The third

    D. The sixth

    E. The fifth

20. In a frequency distribution, the frequency always acts as the

    A. independent variable.

    B. dependent variable.

    C. population.

    D. sampling frame.

    E. standard deviation.

21. Which of the following parameters would we *always* have to treat as a discrete variable in a relevant statistical experiment?

    A. The time required for paint to dry

    B. The temperature of the air

    C. The face of a die that turns up when we toss it once

    D. The speed of a moving object

    E. The distance between two moving objects

22. In a normal distribution, we define the number of standard deviations by which an element falls above or below the mean as the

    A. Z score.

    B. coefficient of variation.

    C. variance.

    D. interquartile range.

    E. central tendency.

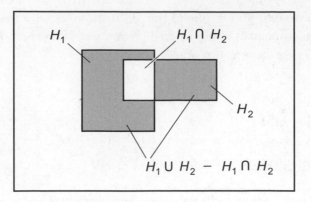

FIGURE EXAM-3 • Illustration for Final Exam Question 23.

23. In the Venn diagram of Fig. Exam-3, the shaded regions portray the union of the two sample spaces H_1 and H_2 for an arbitrarily large set of identical experiments in which we observe outcomes h_1 and h_2, minus the intersection of those same two sample spaces H_1 and H_2. We have

$$p(h_1 \vee h_2) = p(h_1) + p(h_2) - p(h_1 \wedge h_2)$$

Here, the wedge symbol $\vee$ represents the logical "or" operation in the inclusive sense, meaning "either or both," while the inverted V symbol ($\wedge$) stands for the logical connector "and." These outcomes are

A. complementary.

B. contradictory.

C. supplementary.

D. nondisjoint.

E. mutually inclusive.

24. Imagine that you, I, and our mutual friend Joanna plan to toss a pair of six-faced, "unweighted" gambling dice. I suggest that the dice will land showing two to five dots in total. You suggest that the dice will land showing six to nine dots in total. Joanna disagrees with us both, suggesting that the dice will come up showing 10 to 12 dots in total. We all agree to consider my suggestion as a "reference point," and then you and Joanna hope to prove me wrong when we actually throw the dice. My suggestion constitutes

A. an event.

B. a null hypothesis.

C. a model.

D. a one-sided alternative.

E. a two-sided alternative.

25. Which of the following actions *should not* concern us if we want to avoid experimental defect error?

    1. Neglecting factors that can introduce bias

    2. Inappropriate replacement of elements

    3. The use of a sample that's too small

    4. The use of hardware with too much precision

    5. Compensating for factors that have no practical effect

    A. Action 1

    B. Actions 1 and 2

    C. Actions 2 and 3

    D. Action 4

    E. Actions 3 and 5

26. Figure Exam-4 is a generic graph of a normal distribution. At which *percentile boundary point* does *P* lie, *exactly?*

    A. The 69th

    B. The 70th

    C. The 30th

    D. The 31st

    E. None of the above

27. In Fig. Exam-4, at which *decile boundary point* does *P* lie, *exactly?*

    A. The sixth

    B. The seventh

    C. The third

    D. The fourth

    E. None of the above

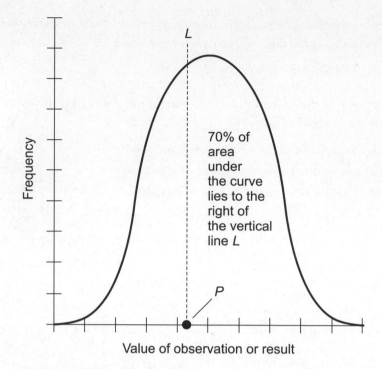

FIGURE EXAM-4 • Illustration for Final Exam Questions 26 through 28.

28. In Fig. Exam-4, at which *quartile boundary point* does *P* lie, *exactly?*
    A. The first
    B. The second
    C. The third
    D. The fourth
    E. None of the above

29. We commit the probability fallacy when we
    A. calculate the probability of an event using unproven methods.
    B. inaccurately calculate the probability of an event.
    C. talk about probability where it plays no legitimate role.
    D. falsify data that we later use to calculate probability values.
    E. accept probability data without sufficient evidence of its truth.

30. Some doubt exists as to whether any human or machine can generate a
    truly random sequence of numbers. However, we can use computers to

generate sequences of numbers that behave "in a random way for most practical purposes." We call such numbers

A. regressive.

B. pseudorandom.

C. irrational.

D. transcendental.

E. indefinite.

31. Let's revisit the scenario of Question 24. You, I, and Joanna toss a pair of six-faced, "unweighted" gambling dice. I suggest that the dice will land showing two to five dots in total. You suggest that it will show six to nine dots in total. Joanna suggests that it will show 10 to 12 dots in total. We all agree to let Joanna's suggestion serve as the null hypothesis, which you and I hope to disprove. My hypothesis alone constitutes a

A. singular event.

B. negative hypothesis.

C. regressive model.

D. one-sided alternative.

E. two-sided alternative.

32. Suppose that we begin with the number 17.498172 and approximate it to two significant figures in steps, as follows:

$$17.498172$$
$$17.49817$$
$$17.4982$$
$$17.498$$
$$17.50$$
$$17.5$$
$$18$$

This process constitutes an example of

A. linear interpolation.

B. regression approximation.

C. rounding.

D. linear extrapolation.

E. truncation.

33. Suppose that we begin with the number 17.498172 and approximate it to two significant figures in steps, as follows:

17.498172
17.49817
17.4981
17.498
17.49
17.4
17

This process constitutes an example of

A. linear interpolation.

B. regression approximation.

C. rounding.

D. linear extrapolation.

E. truncation.

34. Variance and standard deviation both express the extent to which the values in a distribution appear clustered around the

A. median.

B. mean.

C. mode.

D. regression curve.

E. origin.

35. In the normal distribution represented by the bell-shaped curve in Fig. Exam-5, we can be *certain* that the Z score of point *x* is

A. an undefined value.

B. a positive number.

C. zero.

D. a negative number.

E. approximately 1.7 standard deviations.

36. Let's revisit the scenario of Questions 24 and 31. You, I, and Joanna toss a pair of six-faced, "unweighted" gambling dice. I suggest that the dice will land showing two to five dots in total. You suggest that it will show

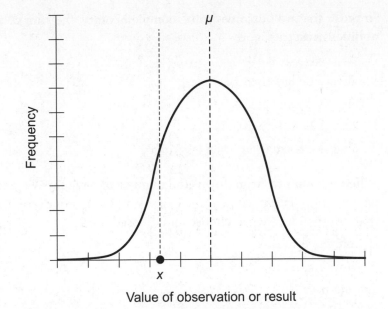

Value of observation or result

FIGURE EXAM-5 • Illustration for Final Exam Question 35.

six to nine dots in total. Joanna suggests that it will show 10 to 12 dots in total. We all agree to let your suggestion serve as the null hypothesis, which Joanna and I hope to disprove. Joanna's suggestion and my suggestion, considered together, constitute a

A. dual event.

B. negative hypothesis.

C. one-sided alternative.

D. two-sided alternative.

E. four-sided alternative.

37. Fill in the blank to make the following sentence true: "In a large data set, _____ percentile points exist."

A. 11

B. 10

C. 101

D. 99

E. 100

38. In order for two outcomes to be complementary, the sum of their probabilities must

    A. precisely equal 1.

    B. always be less than 1.

    C. always exceed 1.

    D. always be less than 0.5.

    E. always exceed 0.5 but be less than 1.

39. When we want to "trim down" the number of elements we examine in a gigantic population while nevertheless obtaining a fair representation of that population, we can choose an unbiased

    A. reference subset.

    B. sampling frame.

    C. data base.

    D. confidence interval.

    E. regression set.

40. As we increase the width of the confidence interval in a normal distribution, the probability that an element selected "at random" will lie within that interval

    A. increases.

    B. decreases.

    C. approaches, but can never exceed, 68%.

    E. approaches, but can never exceed, 95%.

    D. approaches, but can never exceed, 99.7%.

41. A set called P constitutes a *subset* of a set called Q, written $P \subseteq Q$, if and only if, for every possible element z,

    A. $z \in P$ implies that $z \in Q$

    B. $z \in Q$ implies that $z \in P$

    C. $z \notin P$ implies that $z \in Q$

    D. $z \notin Q$ implies that $z \in P$

    E. $z \in P$ if and only if $z \in Q$

42. Fill in the following sentence to make it true: "If we conduct _____, the size of the sample set remains constant, regardless of how many elements it contains at the beginning."

   A. sampling with replacement

   B. regressive sampling

   C. biased sampling

   D. unbiased sampling

   E. pseudorandom sampling

43. Imagine that we conduct an experiment to determine the number of sunny days per month as a function of the rainfall in centimeters per month for 250 different places in the world. We discover that a significant negative correlation exists, and when we calculate it, we get −30%. If we reverse the roles of the independent and dependent variables and calculate the correlation again, it will come out as

   A. −70%.

   B. +70%.

   C. −30%.

   D. +30%.

   E. an undefined value.

44. Figure Exam-6 portrays an example of

   A. a regressive distribution.

   B. a sampling distribution.

   C. a least-squares distribution.

   D. an estimated distribution.

   E. an experimental distribution.

45. In Fig. Exam-6, the vertical line marked X represents

   A. the average standard deviation.

   B. the expected standard deviation.

   C. the standard deviation mode.

   D. the estimated standard deviation.

   E. All of the above

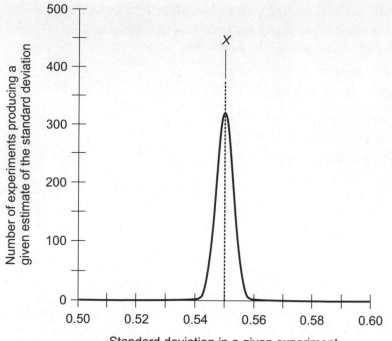

FIGURE EXAM-6 • Illustration for Final Exam Questions 44 and 45.

46. Table Exam-2 shows the results of a hypothetical 10-question quiz given to a class of students. What's the median?

    A. 4

    B. Both 4 and 5 (the distribution is bimedial)

    C. 4.5

    D. 5

    E. We can't define it.

47. Consider two positive integers q and r. We can symbolize the number of different permutations in this situation by writing $_qP_r$ and calculate it with the formula

$$_qP_r = q!/(q-r)!$$

    Based on this formula, how many permutations can we obtain with 12 apples, taken seven at a time in a specific order?

    A. 120

    B. 5040

TABLE EXAM-2 Results of a hypothetical 10-question quiz given to a class of students. This table goes with Final Exam Question 46.	
Test Score	**Absolute Frequency**
0	1
1	5
2	4
3	11
4	16
5	16
6	12
7	7
8	2
9	0
10	0

C. 524,288

D. 1,048,576

E. 3,991,680

48. Imagine three phenomena called *A*, *B*, and *C*, all of which show mutual positive correlation. In other words, *A* positively correlates with *B*, *B* positively correlates with *C*, and *A* positively correlates with *C*. Based on these observations, but before we gather any scientific data (other than the mere existence of the correlations), which of the following statements can we make with absolute confidence?

A. One of the phenomena causes the other two.

B. One pair of the phenomena causes the third.

C. Some external factor causes all three phenomena.

D. None of the above statements A, B, or C can possibly be correct.

E. All of the above statements A, B, and C might be wrong.

49. Once again, imagine three phenomena called *A*, *B*, and *C*, all of which show mutual positive correlation as in the scenario of Question 48. Based on these observations, but before we gather any scientific data (other than the mere existence of the correlations), which, if any, of the following statements A, B, or C constitutes a plausible null hypothesis?

    A. One of the phenomena causes the other two.

    B. One pair of the phenomena causes the third.

    C. Some external factor causes all three phenomena.

    D. Any of the above statements A, B, or C constitutes a plausible null hypothesis.

    E. None of the above A, B, or C constitute a plausible null hypothesis.

50. If a set of numbers has an upper bound, then we can have confidence that it has

    A. a least upper bound.

    B. a greatest upper bound.

    C. a least lower bound.

    D. a greatest lower bound.

    E. All of the above

51. Suppose that we encounter an endless sequence of digits from the set {0, 1, 2, 3, 4, 5, 6, 7, 8, 9}. Somehow we manage to prove that no human or machine can determine the value of any given digit based on the values of previous digits in the sequence. In that case, we can have confidence that the digits

    A. fall within a normal distribution.

    B. fall within a skewed distribution.

    C. occur in truly random fashion.

    D. follow the principle of the least upper bound.

    E. follow the principle of the greatest lower bound.

52. In a normal distribution, the mean appears at the same point where the function

    A. "blows up" toward "infinity."

    B. attains a value of 0.

    C. attains its minimum value.

    D. attains its maximum value.

    E. crosses the independent-variable axis.

TABLE EXAM-3	Results of a hypothetical 10-question quiz given to a class of students. This table goes with Final Exam Question 53.

Test Score	Absolute Frequency
0	0
1	0
2	5
3	7
4	10
5	12
6	17
7	22
8	24
9	18
10	11

53. Table Exam-3 shows the results of a hypothetical 10-question quiz given to a class of students. What's the median?

   A. 6

   B. Both 6 and 7 (the distribution is bimedial)

   C. 6.5

   D. 7

   E. We can't define it.

54. How can we state the law of large numbers?

   A. As the number of events in an experiment increases, the average value of the outcome approaches the theoretical median.

   B. As the number of events in an experiment increases, the average value of the outcome approaches the theoretical mean.

   C. As the number of events in an experiment increases, the average value of the outcome approaches the theoretical mode.

   D. As the number of events in an experiment increases, the average value of the outcome approaches the theoretical standard deviation.

   E. As the number of events in an experiment increases, the average value of the outcome approaches the theoretical variance.

55. Suppose that we administer a test to a class of students and tally up the number of students who got each possible score. We multiply each individual score by its absolute frequency, obtaining a set of products. Then we add up the products and divide the result by the number of papers in the class. This process tells us the

    A. normal distribution of means.

    B. standard deviation.

    C. variance.

    D. population mean.

    E. median.

56. Suppose that we observe two quantifiable phenomena X and Y and, through extensive testing, find that they exhibit a correlation of −68%. In qualitative terms, we can say that

    A. as the value of X gets more positive, the value of Y gets more negative in all cases.

    B. as the value of X gets more positive, the value of Y gets more positive in all cases.

    C. as the value of X gets more positive, the value of Y gets more negative in some, but not all, cases.

    D. as the value of X gets more positive, the value of Y gets more positive in some, but not all, cases.

    E. we've made a mistake somewhere.

57. Someone tells you that all normal distributions have the following three characteristics, where the symbol σ stands for the standard deviation, the symbol μ stands for the mean, and the symbol $\pm$ stands for "plus-or-minus":

    - Approximately 68% of the data points lie within the range $\pm\sigma$ of μ.
    - Approximately 95% of the data points lie within the range $\pm2\sigma$ of μ.
    - Approximately 99.7% of the data points lie within the range $\pm3\sigma$ of μ.

    What, if anything, is wrong with this claim as stated?

    A. The percentage values are too large by a factor of 2.

    B. The percentage values are too large by a factor of the square root of 2.

    C. The percentage values are too small by a factor of 2.

D. The percentage values are too small by a factor of the square root of 2.

E. Nothing is wrong with the claim as stated.

58. **The claim made in Question 57, when corrected (if necessary), is known as the**

A. normal distribution rule.

B. standard distribution rule.

C. continuous distribution rule.

D. empirical rule.

E. normal curve-fitting rule.

59. **Suppose that we upload a gigantic video file to the Internet. We plot the data transfer rate, in megabits per second, as a function of the time in seconds. In this situation, we would most likely consider the data transfer rate as**

A. the range.

B. the dependent variable.

C. a nondecreasing value.

D. the independent variable.

E. a sample space.

60. **As we increase the size of a sample and repeatedly graph the distribution for the population that the sample represents, we can have complete confidence that our estimate of the mean will approach**

A. the true median for the distribution.

B. the true mode for the distribution.

C. the true variance for the distribution.

D. all of the values A, B, and C.

E. none of the values A, B, or C.

61. **Are the variance and the standard deviation directly related in a statistical distribution? If so, how?**

A. Yes, the standard deviation equals the square root of the variance.

B. Yes, the variance equals the square root of the standard deviation.

C. Yes, the standard deviation equals the variance divided by the median.

D. Yes, the variance equals the standard deviation divided by the median.

E. No, they aren't directly related.

62. A description of the set of all possible values that a random variable can attain is called a statistical

    A. distribution.

    B. range.

    C. parameter.

    D. experiment.

    E. sample.

63. Suppose that we observe two quantifiable phenomena X and Y and, through extensive testing, find that they exhibit a correlation of +100%. In qualitative terms, we can say that

    A. as the value of X gets more positive, the value of Y gets more negative in all cases.

    B. as the value of X gets more positive, the value of Y gets more positive in all cases.

    C. as the value of X gets more positive, the value of Y gets more negative in some, but not all, cases.

    D. as the value of X gets more positive, the value of Y gets more positive in some, but not all, cases.

    E. we've made a mistake somewhere.

64. Scatter plots and regression curves, used together, can give us an excellent way to visualize the

    A. variance in a normal distribution.

    B. standard deviation in a normal distribution.

    C. nature and strength of the correlation between two variables.

    D. mean, median, and mode in a discrete distribution.

    E. All of the above

65. Figure Exam-7 shows a scatter plot for 13 values (solid black dots) of an independent variable C and a dependent variable M. Based on the appearance of this graph, we can conclude that the correlation between these two variables is

    A. strong and positive.

    B. weak and positive.

    C. zero or near zero.

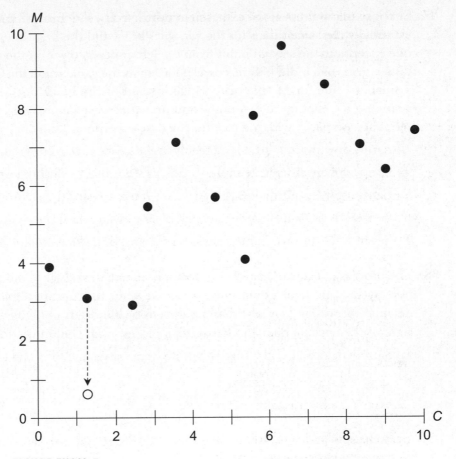

FIGURE EXAM-7 • Illustration for Final Exam Questions 65 through 67.

    D. weak and negative.

    E. strong and negative.

66. **If we use a computer to determine and draw the least-squares line in Fig. Exam-7, we'll see that as we move toward the right on the coordinate grid, the line**

    A. ramps downward steeply.

    B. ramps downward gradually.

    C. appears horizontal or nearly horizontal.

    D. ramps upward gradually.

    E. ramps upward steeply.

67. Suppose that we repeat the experiment portrayed by Fig. Exam-7 and get essentially the same values for the variables as we did the first time, with one exception: the second point from the left appears lower on the coordinate grid than it did before, corresponding to the same value for C but a smaller value for M (the open circle instead of the black dot). If we again use a computer to determine and draw the least-squares line, we'll see that, compared with the plot for the first experiment, the line

   A. ramps downward a little less steeply as we move toward the right.

   B. ramps downward a little more steeply as we move toward the right.

   C. ramps upward a little less steeply as we move toward the right.

   D. ramps upward a little more steeply as we move toward the right.

   E. has moved upward slightly, but hasn't changed in slope at all.

68. The process of *truncation* allows us to approximate the values of numbers portrayed as decimal expansions. When we want to truncate a massive decimal expression (say, one that has dozens or hundreds of digits to the right of the decimal point) to six decimal places, what should we do?

   A. Simply delete all of the single-digit numerals after the fifth one to the right of the decimal point.

   B. Simply delete all of the single-digit numerals after the sixth one to the right of the decimal point.

   C. Simply delete all of the single-digit numerals after the seventh one to the right of the decimal point.

   D. Delete all of the single-digit numerals after the fifth one to the right of the decimal point; then leave the fifth numeral alone if the sixth one was 0 through 4, or increase the fifth one by 1 if the sixth one was 5 through 9.

   E. Delete all of the single-digit numerals after the sixth one to the right of the decimal point. Then leave the sixth numeral alone if the seventh one was 0 through 4, or increase the sixth one by 1 if the seventh one was 5 through 9.

69. What are the real-number solutions to the equation $(x - 3)(x - 7) = 0$?

   A. $x = -3$ and $x = -7$

   B. $x = 3$ and $x = 7$

   C. $x = -10$

   D. $x = 21$

   E. None exist.

70. Which of the following disciplines takes note of the apparent fact that events tend to occur in bunches?

    A. Chaos theory

    B. Correlation theory

    C. Causation theory

    D. Number theory

    E. Regression theory

71. When we create tables or graphs that portray two variable quantities side by side for easy comparison, we make use of

    A. geometric data.

    B. contrasted data.

    C. relational data.

    D. dual-function data.

    E. paired data.

72. Imagine that an election takes place for a seat in the state assembly. Two candidates run for office in our district, one of whom calls himself "liberal" and the other of whom calls herself "conservative." Imagine that the "conservative" candidate gets 54% of the popular vote, thereby winning the election. Suppose that we choose 10 ballots "at random" out of various ballot boxes scattered "randomly" around the district. What's the probability, to the nearest percentage point, that we'll select 10 ballots, all of which indicate votes for the winning candidate?

    A. 27%

    B. 13%

    C. 11%

    D. 5%

    E. Less than 1%

73. The graph of a normal distribution always exhibits bilateral (left-to-right) symmetry with respect to a vertical line corresponding to the

    A. mean.

    B. variance.

    C. standard deviation.

    D. dependent variable.

    E. independent variable.

74. If two sets have *no* elements in common, we call them

    A. null sets.

    B. complementary sets.

    C. supplementary sets.

    D. empty sets.

    E. disjoint sets.

75. We can loosely define the term *mean* for a discrete variable in a distribution as the

    A. value(s) that occur(s) with the highest absolute frequency.

    B. value(s) that occur(s) with the lowest absolute frequency.

    C. sum of all the values, divided by the number of values.

    D. square root of the product of all the values.

    E. point representing the "middle" of the distribution.

76. If a set of numbers has a lower bound, then we can have confidence that the set has

    A. an infimum.

    B. a normal distribution.

    C. a uniform distribution.

    D. a least-squares line.

    E. a regression curve.

77. A sample of a population always constitutes a subset of that population. What do we call a single element of a sample?

    A. A continuous variable

    B. A discrete variable

    C. A parameter

    D. An event

    E. A sample space

78. Which of the following descriptions best defines the term *sampling frame*?

    A. A set of samples from which we compile a population

    B. A set of elements from within a population, from which we choose a sample

C. A set of samples from within a population, all of which intersect to yield a set containing at least one element

D. A set of elements from within a sample, all of which have similar characteristics

E. The process by which we go about obtaining a population from within a sample containing infinitely many elements

79. Imagine a town where the average monthly rainfall is *slightly* greater in the winter than in the summer; the spring and autumn months are wetter than the summer months but drier than the winter months. The average monthly temperature is *slightly* higher in the summer than in the winter on the average; the spring and autumn months are cooler than the summer months but warmer than the winter months. However, occasional temperature anomalies can exist in which a summer month might actually turn out cooler than a spring or autumn month in the same year. In this town, the correlation between average monthly temperature and average monthly precipitation is

A. strongly positive.

B. strongly negative.

C. weakly positive.

D. weakly negative.

E. nonexistent.

80. The 75% confidence interval in a normal distribution spans values within plus or minus

A. more than three, but less than four, standard deviations of the estimate of the mean.

B. more than two, but less than three, standard deviations of the estimate of the mean.

C. more than one, but less than two, standard deviations of the estimate of the mean.

D. less than one standard deviation of the estimate of the mean.

E. some undefinable value, because technically there's no such thing as a 75% confidence interval.

81. Figure Exam-8 is a generic graph of a statistical distribution. The vertical dashed line corresponds to a specific value of the independent variable. Based on the information shown here, we can surmise that $L + R$ is

A. much less than 100%.

B. somewhat less than 100%.

C. equal to 100%.

D. somewhat more than 100%.

E. much more than 100%.

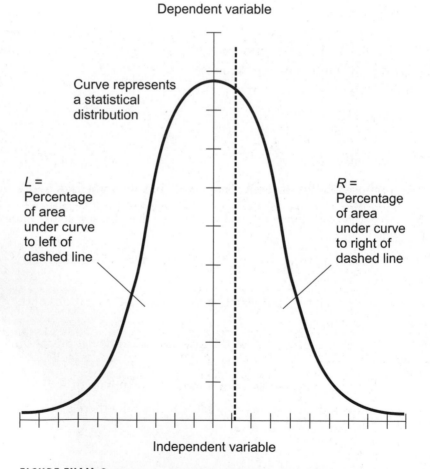

Dependent variable

Curve represents a statistical distribution

$L =$ Percentage of area under curve to left of dashed line

$R =$ Percentage of area under curve to right of dashed line

Independent variable

FIGURE EXAM-8 • Illustration for Final Exam Questions 81 and 82.

82. Suppose that we move the vertical dashed line almost all the way to the left end of the independent-variable axis in Fig. Exam-8. In that case, we can surmise that $L + R$ is

    A. much less than 100%.

    B. somewhat less than 100%.

    C. equal to 100 %.

    D. somewhat more than 100%.

    E. much more than 100%.

83. Consider an equation of the following form where a, b, and c represent real-number constants, x represents a variable, and $a \neq 0$:

    $$ax^2 + bx + c = 0$$

    How many real-number solutions can such an equation have for x?

    A. Exactly one

    B. Exactly two

    C. None, one, or two

    D. One, two, or infinitely many

    E. Either none or else infinitely many

84. Suppose that we observe two quantifiable phenomena X and Y. Through extensive testing, we find that they exhibit a correlation of −142%. In qualitative terms, we can say that

    A. as the value of X gets more positive, the value of Y gets more negative in some, but not most, cases.

    B. as the value of X gets more positive, the value of Y gets more positive in some, but not most, cases.

    C. as the value of X gets more positive, the value of Y gets more negative in most cases.

    D. as the value of X gets more positive, the value of Y gets more positive in most cases.

    E. we've made a mistake somewhere.

85. Figure Exam-9 is a scatter plot showing the results of an experiment in which we've tested the populations of 15 cities located at various altitudes in meters above sea level. In each city, we've determined the percentage of people who drink at least one liter of beer a day (on the average). Based on the appearance of this graph, it appears that

A. a strong negative correlation exists between altitude and the quantity of beer a person drinks daily.

B. a weak negative correlation exists between altitude and the quantity of beer a person drinks daily.

C. little or no correlation exists between altitude and the quantity of beer a person drinks daily.

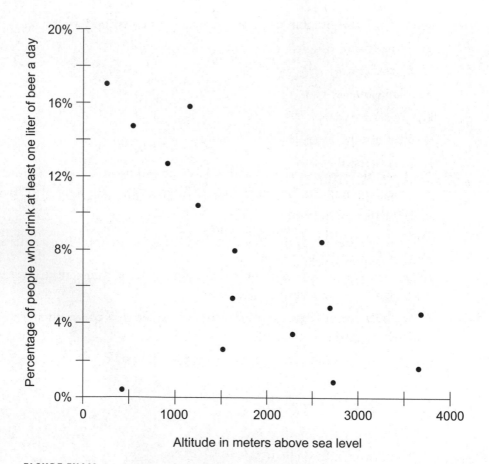

FIGURE EXAM-9 • Illustration for Final Exam Questions 85 and 86.

D. a weak positive correlation exists between altitude and the quantity of beer a person drinks daily.

E. a strong positive correlation exists between altitude and the quantity of beer a person drinks daily.

86. If we use a computer in an effort to find the least-squares line for the scatter plot in Fig. Exam-9, we should expect that our computer-generated line will

A. run horizontally across the coordinate grid.

B. run vertically across the coordinate grid.

C. ramp upward as we move toward the right in the coordinate grid.

D. ramp downward as we move toward the right in the coordinate grid.

E. fail to materialize at all.

87. Which of the following disciplines takes note of the fact that patterns tend to recur or duplicate over diverse time scales?

A. Chaos theory

B. Correlation theory

C. Causation theory

D. Number theory

E. Regression theory

88. Consider the following three statements concerning percentiles, deciles, and quartiles, along with their associated intervals.

- The 100 percentile points divide a ranked data set into 99 intervals as nearly equal-sized as possible.
- The 10 decile points divide a ranked data set into nine intervals as nearly equal-sized as possible.
- The four quartile points divide a ranked set into three intervals as nearly equal-sized as possible.

Which, if any, of these statements is false?

A. None of the statements are false.

B. The first statement is false.

C. The second statement is false.

D. The third statement is false.

E. All three of the statements are false.

89. Imagine that we conduct a statistical experiment among people who live in cities and towns with populations exceeding 1000. We discover that as the population of the city or town increases, in general, the number of people who wear leather boots at their daily jobs decreases. What sort of correlation is this?

    A. Negative

    B. Positive

    C. Zero

    D. Normal

    E. Uniform

90. Which of the following cause/effect relationships does the result of the foregoing statistical experiment (described in Question 89) logically imply all by itself?

    A. Life in a small town causes people to wear boots more often than does life in a big city.

    B. A dislike for boots causes people to move to big cities.

    C. A fondness for boots causes people to move to small towns.

    D. Some unknown factor causes people to either like boots and move to small towns, or to dislike boots and move to big cities.

    E. None of the above

91. As the value a positive integer n increases one by one, the value of $n!$ (n factorial) increases with great rapidity. When n exceeds 50 or so, we can approximate $n!$ using the formula

$$n! \approx n^n/e^n$$

    What does e represent here?

    A. The ratio of a square's diagonal to one of its edges, or roughly 1.41421

    B. The ratio of a cube's diagonal to one of its edges, or roughly 1.73205

    C. The natural logarithm base, or roughly 2.71828

    D. The natural logarithm of 10, or roughly 2.30259

    E. The common logarithm of 2, or roughly 0.30103

92. In a normal distribution, we define the value of the third quartile point minus the value of the first quartile point as the interquartile

A. differential.

B. variance.

C. deviation.

D. range.

E. dispersion.

93. Consider two *mutually exclusive* sample spaces H_1 and H_2 for an arbitrarily large set of identical experiments in which we observe two different outcomes: one called h_1 that occurs with probability $p(h_1)$, and the other called h_2 that occurs with probability $p(h_2)$. In this situation, what's the probability of either outcome (but not both) occurring in a single experiment?

A. $p(h_1)p(h_2)$

B. $[p(h_1)p(h_2)]^{1/2}$

C. $[p(h_1) + p(h_2)]/2$

D. $p(h_1) + p(h_2)$

E. $p(h_1)p(h_2)/[p(h_1) + p(h_2)]$

94. Statisticians quantify the extent of the correlation between two variables by means of a technique called

A. pseudorandom testing.

B. regression testing.

C. prediction testing.

D. geometric testing.

E. calculus testing.

95. Suppose that we plot the graph of a density function for a continuous random variable, obtaining the result shown in Fig. Exam-10. For values of the random variable between 13.00 and 14.50, this graph portrays a

A. normal distribution.

B. least-squares distribution.

C. bimodal distribution.

D. uniform distribution.

E. square distribution.

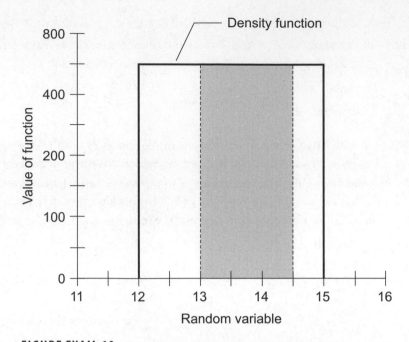

FIGURE EXAM-10 • Illustration for Final Exam Questions 95 and 96.

96. If we restrict our attention to values of the random variable exactly between 13.00 and 14.50 in Fig. Exam-10 (shaded region), what's the probability, to the nearest hundredth of a percent, that a "randomly" selected value of the random variable will produce a function value of 600?

   A. 25%

   B. 33%

   C. 50%

   D. 67%

   E. 100%

97. According to the empirical rule for normal distributions, approximately 68% of the data points have Z scores

   A. between −1 and +1.

   B. between −2 and +2.

   C. between 0 and 1.

   D. between −1 and 0.

   E. greater than +1 or smaller than −1.

98. **A mathematical relation constitutes a function if and only if**

    A. no element in the set of its independent-variable values has a correspondent in the set of dependent-variable values.

    B. every element in the set of its independent-variable values has exactly one correspondent in the set of dependent-variable values, and vice versa.

    C. every element in the set of its independent-variable values has at least one correspondent in the set of dependent-variable values.

    D. every element in the set of its independent-variable values has at most one correspondent in the set of dependent-variable values.

    E. no element in the set of its dependent-variable values has a correspondent in the set of independent-variable values.

99. **We can loosely define the term *median* for a discrete variable in a distribution as the**

    A. value(s) that occur(s) with the highest absolute frequency.

    B. value(s) that occur(s) with the lowest absolute frequency.

    C. sum of all the values, divided by the number of values.

    D. square root of the product of all the values.

    E. point representing the "middle" of the distribution.

100. **We know *for certain* that we have an irrational number when we *cannot* express it as**

    A. the ratio of an integer to a nonzero natural number.

    B. the ratio of two positive integers.

    C. a decimal expression of finite length.

    D. an endless, repeating decimal.

    E. a whole number.

$q = 4$

$_4P_3 = 4!(4-3)!$

$r = 3$

$= 4!/1!$

$= 4!$

$= 24$

Answers to Quizzes, Tests, and Final Exam

Chapter 1	Chapter 3	Test: Part I	
1. C	1. B	1. C	23. A
2. D	2. B	2. B	24. D
3. D	3. D	3. D	25. C
4. B	4. B	4. E	26. E
5. C	5. C	5. E	27. B
6. B	6. B	6. E	28. C
7. A	7. A	7. C	29. D
8. A	8. C	8. E	30. B
9. B	9. A	9. A	31. A
10. A	10. A	10. C	32. E
		11. E	33. B
Chapter 2	Chapter 4	12. B	34. C
1. B	1. C	13. D	35. B
2. B	2. D	14. A	36. C
3. D	3. D	15. D	37. E
4. A	4. D	16. A	38. E
5. D	5. D	17. C	39. B
6. C	6. B	18. E	40. C
7. A	7. C	19. A	41. B
8. B	8. A	20. C	42. C
9. C	9. B	21. C	43. D
10. C	10. A	22. A	44. A
			45. C

46. E
47. A
48. D
49. D
50. B
51. E
52. B
53. A
54. D
55. C
56. A
57. C
58. D
59. D
60. C

Chapter 5
1. D
2. A
3. D
4. C
5. D
6. C
7. A
8. A
9. D
10. B

Chapter 6
1. D
2. B
3. A
4. B
5. A
6. A
7. C
8. C
9. D
10. A

Chapter 7
1. C
2. A
3. A
4. B
5. B
6. D
7. C
8. B
9. D
10. B

Chapter 8
1. B
2. D
3. C
4. D
5. B
6. D
7. A
8. D
9. C
10. A

Test: Part II
1. C
2. B
3. E
4. D
5. A
6. B
7. D
8. E
9. C
10. C
11. A
12. E
13. B
14. D

15. A
16. D
17. D
18. C
19. C
20. A
21. E
22. B
23. A
24. B
25. C
26. A
27. D
28. E
29. B
30. E
31. A
32. D
33. B
34. D
35. E
36. A
37. E
38. D
39. B
40. A
41. B
42. B
43. B
44. D
45. C
46. A
47. A
48. B
49. E
50. D
51. B
52. E

53. B
54. D
55. C
56. C
57. E
58. D
59. E
60. A

Final Exam
1. C
2. B
3. E
4. E
5. D
6. D
7. B
8. A
9. E
10. E
11. A
12. C
13. A
14. B
15. B
16. A
17. B
18. D
19. E
20. B
21. C
22. A
23. D
24. B
25. D
26. C
27. C
28. E

29. C	47. E	65. B	83. C
30. B	48. E	66. D	84. E
31. D	49. D	67. D	85. B
32. C	50. A	68. B	86. D
33. E	51. C	69. B	87. A
34. B	52. D	70. A	88. E
35. D	53. D	71. E	89. A
36. D	54. B	72. E	90. E
37. D	55. D	73. A	91. C
38. A	56. C	74. E	92. D
39. B	57. E	75. C	93. D
40. A	58. D	76. A	94. B
41. A	59. B	77. D	95. D
42. A	60. E	78. B	96. E
43. C	61. A	79. D	97. A
44. B	62. A	80. C	98. D
45. A	63. B	81. C	99. E
46. C	64. C	82. C	100. A

Suggested Additional Reading

Downing, Douglas and Clark, Jeffrey, *Statistics the Easy Way*, 3rd ed. Barron's Educational Series, 1997.

Gibilisco, Stan, *Technical Math Demystified*. McGraw-Hill, 2006.

Gibilisco, Stan and Crowhurst, Norman, *Mastering Technical Mathematics*, 3rd ed. McGraw-Hill, 2007.

Graham, Alan, *Teach Yourself Statistics*, 3rd ed. McGraw-Hill, 2008.

Jaisingh, Lloyd, *Statistics for the Utterly Confused*, 2nd ed. McGraw-Hill, 2005.

Kemp, Steven and Kemp, Sid, *Business Statistics Demystified*. McGraw-Hill, 2004.

Koosis, Donald, *Statistics: A Self-Teaching Guide*, 4th ed. John Wiley & Sons, Inc., 1997.

Levin, Richard I. and Rubin, David S., *Statistics for Management*, 7th ed. Prentice-Hall, Inc., 1997.

Moore, David S. and Notz, William I., *Statistics: Concepts and Controversies*, 7th ed. W. H. Freeman & Co., 2008.

Schiller, John, Srinivasan, R. Alu, and Spiegel, Murray, *Probability and Statistics*, 3rd ed. Schaum's Outline Series, McGraw-Hill, 2008.

Spiegel, Murray R. and Stephens, Larry J., *Statistics*, 4th ed. Schaum's Outline Series, McGraw-Hill, 2007.

Stephens, Larry J., *Advanced Statistics Demystified*. McGraw-Hill, 2004.

Stephens, Larry J., *Beginning Statistics*, 2nd ed. Schaum's Outline Series, McGraw-Hill, 2009.

Stephens, Larry J., *Engineering Statistics Demystified*. McGraw-Hill, 2004.

Index

A

absolute frequency, 45, 53–54, 62–64
alternative
  one-sided, 194
  two-sided, 194
alternative hypothesis, 193–197, 202–204, 296
analog display, 168
arithmetic mean, 18–19
assumptions, 190–196
average, 18–19

B

bar graph, 25, 27–28, 206–208, 262–263
binary digital process, 168
blood pressure, 108–109
bunching of events, 238
butterfly effect, 245–246

C

c% confidence interval, 180–182
causation versus correlation, 230–231
cause and effect, 229–237, 299–302
census, 44
Central Limit Theorem, 175–176
central tendency, 59
chaos theory, 236, 238–250
choosing sampling frames, 159–161
coefficient of variation, 131–133
coincidence, 235–236, 238–239
coincident sets, 7
combinations, 92–93

complementary outcomes, 81–83

confidence interval
  68%, 178, 184–185
  95%, 178–180, 184–185
  99.7%, 180–181, 184–185
  c%, 180–182
confidence intervals and forecasting, 197–198
continuous distribution, 46–48
continuous variable, 41–42
correlation, 34–35, 222–229, 239–240, 295–302
correlation range, 222–223
cumulative absolute frequency, 53–54, 62–63
cumulative relative frequency, 54–55
curve, smooth, 24
curve fitting, 30–31
cycles per second, 41

D

damped oscillation, 248–249
data intervals
  by element quantity, 120–123, 281–287
  fixed, 124–131, 281–287
  range of, 131
deciles, 118–120, 284
decimal expansions, 16
decimal form, 16
decimal point, 16
density function, 93–96
dependent variable, 8–10
descriptive measures, 107–138
dimensionless quantity, 132
discrete distribution, 46–48
discrete values, 24

DeMYSTiFieD®

Hard stuff made easy

The DeMYSTiFieD series helps students master complex and difficult subjects. Each book is filled with chapter quizzes, final exams, and user friendly content. Whether you want to master Spanish or get an A in Chemistry, DeMYSTiFieD will untangle confusing subjects, and make the hard stuff understandable.

PRE-ALGEBRA DeMYSTiFied, 2e
Allan G. Bluman
ISBN-13: 978-0-07-174252-8 • $20.00

ALGEBRA DeMYSTiFied, 2e
Rhonda Huettenmueller
ISBN-13: 978-0-07-174361-7 • $20.00

CALCULUS DeMYSTiFied, 2e
Steven G. Krantz
ISBN-13: 978-0-07-174363-1 • $20.00

PHYSICS DeMYSTiFied, 2e
Stan Gibilisco
ISBN-13: 978-0-07-174450-8 • $20.00

Learn more. **Mc Graw Hill** Do more.